Model Reference Adaptive Control

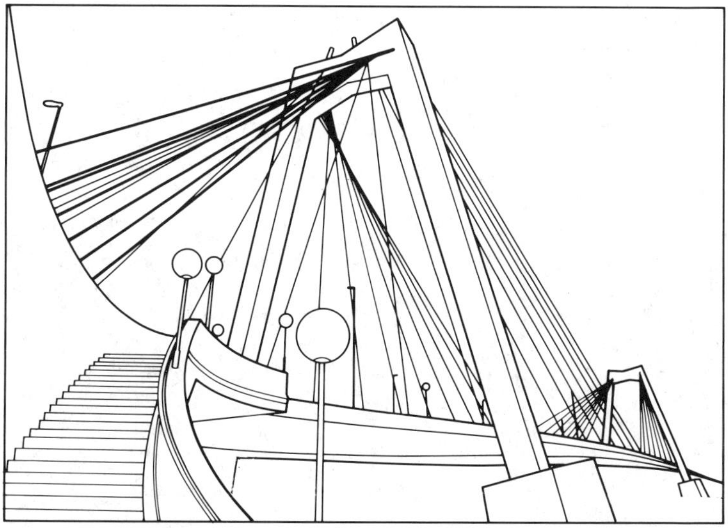

This drawing shows the Willemsbrug, Rotterdam, symbolizing a bridge over the gap between theory and practice in adaptive control. Although the actual bridge is not so large, the angle of composition makes it look huge, in the same way that many people feel that the gap between theory and practice is extremely difficult to bridge. Sometimes reality is more favourable than its image . . .

Prentice Hall International
Series in Systems and Control Engineering

M. J. Grimble, Series Editor

BENNETT, S., *Real-time Computer Control: an Introduction*
BITMEAD, R. R., GEVERS, M. and WERTZ, V., *Adaptive Optimal Control*
CEGRELL, T., *Power Systems Control*
COOK, P. A., *Nonlinear Dynamical Systems*
ISERMANN, R., LACHMANN, K. H, and MATKO, D., *Adaptive Control Systems*
KUCERA, V., *Analysis and Design of Discrete Linear Control Systems*
LUNZE, J., *Feedback Control of Large-Scale Systems*
LUNZE, J., *Robust Multivariable Feedback Control*
McLEAN, D., *Automatic Flight Control Systems*
PATTON, R., CLARK, R. N. and FRANK, P. M. (editors), *Fault Diagnosis in Dynamic Systems*
PETKOV, P. H., CHRISTOV, N. D. and KONSTANTINOV, M. M., *Computational Methods for Linear Control Systems*
SÖDERSTROM, T. and STOICA, P., *System Identification*
WARWICK, K., *Control Systems: an Introduction*
WATANABE, K., *Adaptive Estimation and Control*
STOORVOGEL, A., *The H ∞ Control Problem*
WILLIAMSON, D., *Digital Control and Instrumentation*

MODEL REFERENCE ADAPTIVE CONTROL

From Theory to Practice

Hans Butler

Department of Electrical Engineering,
Delft University of Technology,
The Netherlands

Prentice Hall
New York London Toronto Sydney Tokyo Singapore

First published 1992 by
Prentice Hall International (UK) Ltd
Campus 400, Maylands Avenue,
Hemel Hempstead, Hertfordshire,
HP2 7EZ
A division of
Simon & Schuster International Group

Printed and bound in Great Britain
at the University Press, Cambridge

Library of Congress Cataloging-in-Publication Data

Butler, Hans
Model reference adaptive control: From theory to practice / Hans Butler.
p. cm. – (Prentice Hall international series in systems and control engineering)
Includes bibliographical references and index.
ISBN 0-13-588286-9
1. Adaptive control systems. I. Title II. Series.
TJ216.B87 1992
629.8′36–dc20 91-43416
CIP

British Library Cataloguing in Publication Data

Butler, Hans
Model reference adaptive control: From theory to practice. – (Prentice Hall International series in systems and control engineering)
I. Title II. Series
629.8

ISBN 0-13-588286-9

1 2 3 4 5 96 95 94 93 92

To my Parents

Contents

Preface **xi**

Glossary **xiii**

1 Introduction to Model Reference Adaptive Control **1**

1.1 Adaptive control: why and how? 2

1.2 Model reference adaptive systems 4

1.3 Adaptive controller design 7

1.3.1 The process 7

1.3.2 The primary controller 7

1.3.3 The reference model 8

1.3.4 Derivation of the adaptive laws 9

1.3.5 Averaging techniques 22

1.4 MRAC and other methods 26

1.5 Outline of the chapters 28

Problems 32

2 Continuous-Time MRAC **34**
2.1 MRAC with complete state information 35
2.1.1 Primary controller structure 35
2.1.2 Derivation of the adaptive laws 38
2.2 Extension to the multi-input, multi-output case 44
2.2.1 Process and model description 44
2.2.2 Primary controller structure 45
2.2.3 Derivation of the adaptive laws 46
2.3 MRAC using output feedback 50
2.4 The augmented error method 53
2.4.1 Primary controller structure 53
2.4.2 Derivation of the adaptive laws 55
2.5 An alternative primary controller structure 62
2.6 Simulation examples 68
2.7 Tuning aspects 73
2.8 Summary 75
Problems 76

3 Robustness Topics in Adaptive Control **77**
3.1 Introduction: what is robustness? 78
3.2 Robustness in model reference adaptive control 79
3.2.1 Analysis tools 80
3.2.2 External disturbances 85
3.2.3 Other disturbances 89
3.3 Robustness-improving mechanisms 90
3.3.1 Modifications of the adaptive law 90
3.3.2 Deterministic input disturbances 96
3.3.3 Making use of an orthogonal error signal 101
3.4 Summary 106
Problems 107

4 Discrete-Time MRAC **109**
4.1 Introduction 110
4.2 Going from continuous to discrete MRAC 111
4.2.1 Lyapunov's method 111
4.2.2 Hyperstability 112
4.2.3 One-sample delay 113
4.3 The discrete-time augmented error method 116
4.4 An example using discrete MRAC 118
4.4.1 The auxiliary signal generators 118
4.4.2 The filters L^{-1} 121

4.4.3 Least-squares adaptation 122
4.5 Least-squares adaptation methods 123
4.5.1 A modification of hyperstability theory 123
4.5.2 Least-squares parameter updating 127
4.6 Independent tracking and regulation 128
4.6.1 The primary ITR controller 128
4.6.2 Adaptive ITR 131
4.6.3 General discrete MRAC 133
4.7 Deadbeat model reference adaptive control 134
4.8 Practical results with a water-level system 139
4.9 Summary 146
Problems 147

5 Structured Unmodelled Dynamics in MRAC 148
5.1 Instability caused by unmodelled dynamics 149
5.2 Robustness in the presence of unmodelled dynamics 151
5.3 Reference model decomposition 155
5.3.1 Introduction 156
5.3.2 Basic decomposition idea 157
5.3.3 Effects of decomposition on the error equation 159
5.3.4 A first example 164
5.3.5 A second example 169
5.3.6 Decomposition design steps 174
5.4 Application to a gantry crane scale model 175
5.4.1 Process description 175
5.4.2 Mathematical model of the crane 179
5.4.3 The adaptive controller 182
5.4.4 Practical results 183
5.5 Summary 187
Problems 188

6 Adaptive Model Adjustment: an Application 190
6.1 Introduction 191
6.2 The direct-drive motor 192
6.2.1 Introduction 192
6.2.2 Direct-drive motor model 193
6.2.3 Current control 194
6.3 Position control of the motor 195
6.4 The reference model 198
6.5 Experimental results 202
6.6 Summary 208

Problem 209

7 Direct Model Adjustment: an Application **210**
7.1 Problem statement 211
7.2 An example 213
7.3 Application to a helicopter propeller setup 217
7.3.1 Mathematical model of the propeller setup 219
7.3.2 The primary controller 222
7.3.3 The adaptive controller and the reference model 223
7.3.4 Experimental results 224
7.4 Summary 229
Problems 229

8 Epilogue **230**

A Stability theory **236**
A.1 Lyapunov stability 236
A.2 Positivity and passivity 239

B Answers to Problems **241**

Bibliography **253**

Index **259**

Preface

Adaptive control frightens many people. Partly, this is caused by the large amount of mathematical expertise people feel is necessary to understand adaptive control. Fortunately, this feeling is wrong. Adaptive control schemes are mainly different from linear control systems in that they are inherently nonlinear. The nonlinearity in adaptive control schemes is, however, of a relatively simple type, and many theoretical aspects of adaptive control have a linear counterpart. A basic knowledge of control theory is sufficient to grasp the main ideas behind adaptive control and to understand the essence of its theory.

This book is meant to bring adaptive control theory closer to practice. For this purpose you will mainly find *explanations* of adaptive control theory rather than the theory itself. Several laboratory experiments serve as illustrations. As a consequence, the theory you *will* find in this book cannot be very elaborate. However, many references to literature on the more profound theoretical issues are given for further study.

This book is mainly intended for two groups of people. First, it can serve the purpose for research workers and control engineers in industry who want to become familiar with adaptive control. This book covers many practical issues in adaptive control, giving the reader a sound introduction to practical application. Second, this book can be used as an introduction to adaptive control at various levels in universities and polytechnics, for which purpose end-of-chapter exercises and their solutions are included. Although mathematical details are often omitted, by studying this book a thorough *understanding* of adaptive control can be gained.

This book can be divided into two parts. The first part, consisting of the first four chapters, explains well-established subjects in adaptive control. The second part presents solutions to some of the problems that occur if the theoretically required assumptions are not met. This latter part sometimes has a heuristic character, but makes much use of the concepts established in the first part of the book.

Acknowledgements

Writing a book like this requires the support of many people. In many discussions on a variety of subjects, Ger Honderd, Job van Amerongen, Paul van den Bosch and Henk Verbruggen have all contributed a great deal to this book. In many discussions Ronald Soeterboek tried to convince me of the superiority of predictive control over model reference adaptive control, and without much success I tried to put him back on to the right track. These discussions are typical examples of how adherents of different strategies can learn from each other. I also want to express my gratitude to Jacques Richalet, Salwa Abu el Ata and Laurent Delineau at ADERSA, France, for showing me the inside of PFC and introducing the decomposition philosophy in predictive control to me. My colleagues at ASM Lithography should be credited for convincing me that high technology need not necessarily be adaptive. Rajamani Doraiswami has taught me many invaluable aspects of disturbance decoupling and compensation. Gregory Thé did a lot of research on discrete forms of model decomposition. Rogier van de Ree deserves all praise for the original photograph of the Willemsbrug. The help and support of all these people made writing this book possible and pleasant.

Hans Butler

Glossary

This list contains the most important symbols and abbreviations used in this book.

Symbols:

$\boldsymbol{x}_p$: process state vector

y_p : process output

y_p° : *a posteriori* process output (in discrete MRAC)

$\boldsymbol{x}_m$: reference model state vector

y_m : reference model output

$\boldsymbol{e}$: process-model state error

e_1 : process-model output error

e_1° : *a posteriori* process-model error (in discrete MRAC)

ϵ : error signal as used in the adaptation

v : error-augmenting signal

u : process input

r : reference signal

$\boldsymbol{\omega}$: signal vector

$\boldsymbol{\omega}^*$: signal vector in converged state

$\boldsymbol{\xi}$: filtered signal vector

$\boldsymbol{\xi}^*$: filtered signal vector in converged state

L^{-1} : filter to produce $\boldsymbol{\xi}$ from $\boldsymbol{\omega}$

W_p : process transfer function

$\overline{W_p}$: process transfer function including unmodelled dynamics

$\widetilde{W_p}$: unmodelled dynamics in process transfer function

W_m : reference model transfer function

$\overline{W_m}$: output of $\overline{W_p}$ for $\boldsymbol{\theta} = \boldsymbol{\theta}^*$

$\widetilde{W_m}$: $\overline{W_m} - W_m$

W_c : closed-loop process transfer function

$\boldsymbol{\theta}$: primary controller parameter vector

$\boldsymbol{\theta}^*$: parameter vector for which $W_c = W_m$

$\boldsymbol{\phi}$: parameter error vector $\boldsymbol{\theta} - \boldsymbol{\theta}^*$

$\boldsymbol{\Gamma}$: adaptation gain matrix

V : Lyapunov function

μ_0 : degree of persistent excitation

ν : output disturbance

ν_0 : upper bound on $|\nu|$

σ : σ-modification factor

γ : γ-modification factor

$L(\boldsymbol{\Lambda})$: class of superstrictly positive realness

$N(\boldsymbol{\Pi})$: class of passivity

$T_1(s)$: decomposition feedforward polynomial

$T_2(s)$: decomposition correction polynomial

$N(s)$: decomposition denominator polynomial

y_m^* : decomposition model output for $\boldsymbol{\theta} = \boldsymbol{\theta}^*$

Abbreviations:

AEM: augmented error method

ASG: auxiliary signal generator

ITR: independent tracking and regulation

LSQ: least-squares

MSTC: minimum-settling time control

PE: persistently exciting

RFRC: ripple-free response control

SPR: strictly positive real

STC: self-tuning control

SVF: state-variable filter

UAS: uniformly asymptotically stable

A word on notation:

- Boldface upper-case characters denote matrices.
- Boldface lower-case characters denote vectors.
- Although not formally correct, time-domain signals and Laplace- or z-transformed transfer functions are intermingled. For example, $W_m(\boldsymbol{\phi}^T\boldsymbol{\omega})$ means: the output of a transfer function $W_m(s)$ or $W_m(z)$, which has an input $(\boldsymbol{\phi}^T\boldsymbol{\omega})(t)$. This is done to improve legibility.
- In denoting transfer functions or polynomials in s or z, the 's' or 'z' is usually not printed to improve legibility. Which of the two is applicable usually follows from the text. In all chapters except chapter 4, transfer functions are in 's' unless otherwise stated. In chapter 4, transfer functions are in 'z' unless otherwise stated.

1

Introduction to Model Reference Adaptive Control

Adaptive control is one of those research topics that have received much attention from the systems and control theory and engineering societies, but even so it has always been a controversial one. In almost any book on adaptive control you will find the statement 'don't use adaptive control unless it's absolutely necessary' (and so you have it in this one). This restraint is usually caused by the impression many people have that 'adaptive control is complex' or 'adaptive control will cause you more problems than it solves'. Unfortunately, both statements have some truth in them. While much research in adaptive control is very mathematical, it is sometimes difficult to transform a theoretically correct adaptive controller into a practically operable one. Nevertheless, many theoretical developments in the last decades have brought adaptive control theory much closer to practice, which has, however, been successfully hidden by the control theory community. Many theoretical aspects of adaptive control are very simple once you understand them. This chapter will give you a proper understanding of the theories that are central to adaptive controller design, without bothering you with too much detail.

1.1 Adaptive control: why and how?

During the development of modern control theory, it was realized that a fixed controller cannot provide acceptable system behaviour in all situations. Particularly if the process to be controlled has unknown or time-varying parameters, the design of a fixed controller that always satisfies the desired specifications is not straightforward. In the late 1950s this observation led to a deep interest in adaptive algorithms, starting with the development of the gain-scheduling technique, which can be applied if the process behaviour depends in a known way on some external, measurable condition. For example, the reaction of an aircraft to the pilot's commands is directly dependent on the craft's altitude. Because the craft's behaviour as a function of its altitude is accurately known, it is possible to design a controller for every possible altitude such that the closed loop always satisfies the specifications. Through altitude measurement the controller can be adjusted continuously.

In this specific example, a link is necessary that connects an external condition to the process behaviour. If the process output has an effect on the 'external' condition, an obscure feedback loop arises which may violate the stability properties. In addition, if the process changes due to effects other than the external condition, the gain scheduling does not react to the change. While the system does not respond to not-accounted-for changes, whether or not the gain-scheduling technique should be called 'adaptive' is still a matter of taste.

A possible definition of adaptive control is 'a system that adapts itself to changes in the process'. Another definition which is often used but is probably too vague to be useful is 'a system which is designed from an adaptive point of view'. A more useful one is 'a system that consists of a *primary feedback* that takes care of process signal variations and a *secondary feedback* that deals with process parameter changes'. In this definition, the primary feedback is used as in nonadaptive control, and the secondary feedback makes the system adaptive. From this definition it is clear that process parameter variations give rise to adaptation of the system, which is not the case in the gain-scheduling technique. The aim of reacting to parameter changes is to attempt to maintain a high system performance, even if the process parameters are unknown or varying.

Adaptive control schemes can be categorized in many ways. One useful way divides the class of *direct* adaptive schemes from *indirect* ones. A diagram of an indirect adaptive controller is depicted in figure 1.1, which shows that in addition to the primary feedback loop, incorporating the controller, an 'estimation' block is present which estimates the process parameters. These parameters are used in the block 'design', which calculates the controller parameters such that the closed loop of controller and process satisfies the performance specifications.

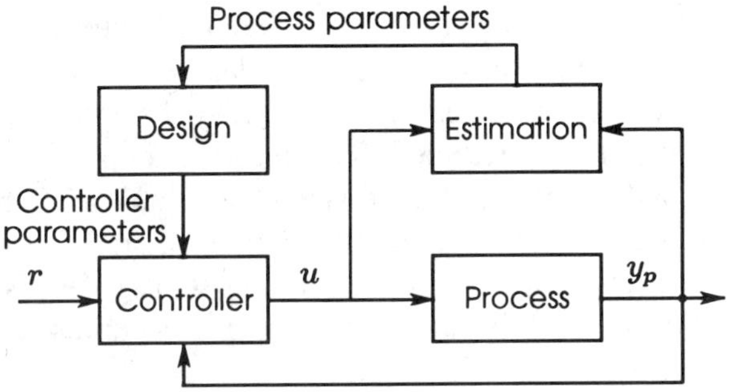

Figure 1.1 An indirect adaptive controller

These specifications are usually formulated as a 'criterion function' denoting the distance between the desired closed-loop behaviour and the actual behaviour. The controller is tuned such that the closed-loop behaviour is optimal in the sense of the criterion function, and hence the block 'design' usually includes a minimization algorithm. Of course, the result of this procedure depends largely upon the correctness of the parameter estimation.

The direct adaptation scheme directly follows from figure 1.1 by deleting the block 'design'. The estimation now produces the controller parameters directly (requiring rewriting of the estimation procedure). Both the indirect and the direct adaptation schemes have their advantages and disadvantages. An advantage of the indirect scheme is the freedom with which the design procedure can be selected. For example, a complex predictive controller can be calculated in the design block (Soeterboek, 1992). A secondary advantage is the availability of the process parameter estimates, which may give the operator a hint of the current state of the process and of possible malfunctions. However, most indirect adaptive controllers do not allow for a theoretical analysis, for the simple reason that the system becomes too complex to analyze. Direct schemes, however, are usually much simpler and therefore do allow for a stability proof, for example. Many theoretical results apply only to this class of adaptive systems.

This analysis problem indicates why adaptive controllers are sometimes still considered 'tricky': the estimation scheme, whether direct or indirect, adds extra dynamics to the control system. These dynamics are always nonlinear, and hence complicate the analysis problem. Model reference adaptive control, as studied in this book, is a typical example of direct adaptation. In many texts the terms 'model reference adaptive control' and 'direct adaptive control' are used as synonyms. However, model reference adaptive controllers exist that are forms of indirect adaptation (see chapter 4). A more thorough attempt to define model reference adaptive control is made later in this chapter.

This chapter first introduces the model reference approach to adaptive controller design, being among the first-proposed adaptive control strategies. Next, several important techniques that are used in MRAC design, in particular in the adaptive law derivations, are presented. This presentation is intended to give the reader an intuitive understanding of the mathematical tools available, without going into too much detail. After that, an overview of the relation that MRAC has with other control methods will be presented, and some recent issues in MRAC research are mentioned. Finally, the organization and purpose of this book are indicated.

1.2 Model reference adaptive systems

The model reference adaptive control technique was first introduced by Whitacker in 1958. The most popular form of a model reference adaptive control (MRAC) scheme is shown in figure 1.2. In this scheme, it can be seen that a *primary*

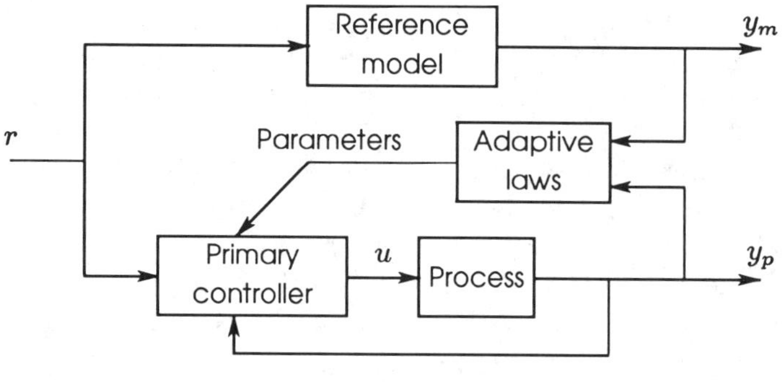

Figure 1.2 General parallel MRAC scheme

controller is used to obtain suitable closed-loop behaviour, as in non-adaptive control schemes. However, because the process parameters are unknown or may vary with time, a fixed parameter setting for the primary controller, such that the closed-loop behaviour is acceptable under all circumstances, cannot be found. In the MRAC technique, the desired process response to a command signal is specified by means of a parametrically defined *reference model*. An *adaptation mechanism* keeps track of the process output y_p and the model output y_m and calculates a suitable parameter setting such that the difference between these outputs tends to zero. In addition to the process output y_p, the process state x_p, if available, and the process input u or the reference signal r may be used by the adaptation mechanism. A characteristic of most (but not all) MRAC schemes is the *direct* adaptation, without an explicit parameter estimation part.

Figure 1.2 shows that an MRAC system consists of the two feedback loops mentioned in section 1.1: an inner loop containing the primary controller, and an outer loop consisting of the adaptation mechanism. Usually, the primary feedback loop operates at a higher speed than the adaptation loop, and thus the process parameters are assumed to vary more slowly than the process states.

Figure 1.2 also shows that the reference model is placed *in parallel* with the process. This is not the case in all MRAC schemes; there are many alternatives, of which the series-type MRAC is the most well known. In this scheme, the reference model is placed in cascade with the process and can be regarded as a trajectory generator. While the closed loop cannot usually follow a stepwise set-point change directly, the reference model output is much smoother and therefore specifies an achievable process response. The series structure has several advantages, such as a simple solution for process input saturation (van Amerongen, 1982), and the possibility of introducing high gains in the primary feedback loop. In other adaptive control schemes, as in self-tuning control or predictive adaptive control, a series-type reference model is also present, and is then usually called a *reference generator*. In addition to the series-type MRAC, there are combinations of a parallel and a series configuration which are called *series–parallel*. In these schemes the reference model is divided into two parts: the first part is placed in series with the process, and the second part in parallel. These schemes are mainly used in identification where the *model* is adjusted instead of the controller parameters. Passenier (1989) describes the specific advantages and disadvantages of the various structures when used in identification. Several other, more complex, structures can be used with more than one adjustable system or in a combination of direct and indirect adaptive algorithms. This book will primarily deal with parallel MRAC.

The MRAC approach has evolved from deterministic servo problems and is still used mainly for this type of application (see, for a broad but shallow overview, Butler, 1990). However, some modern discrete-time MRAC algorithms allow separate specification of the tracking and regulation requirements. These algorithms have a strong resemblance to self-tuning schemes, and some can be shown to be equivalent.

An important issue in MRAC is the design of the adaptive laws. The first adaptive law designs made use of sensitivity models, and later the stability theory of Lyapunov, and Popov's hyperstability theory, served as standard design methods, yielding a guaranteed stable adaptive system. These approaches are described in section 1.3.

Finding a proper definition of 'model reference adaptive control' is not easy. Although the ingredients *primary controller*, *reference model* and *adaptive laws* are found in every MRAC scheme, these elements are also usually present in other adaptive methods, although they may be incognito. In this book, a broad definition is used: if the above-mentioned elements are present under their real names the system is considered to be an MRAC system. This 'definition' allows for a large class of systems to be called MRAC. For example, in the literature an indirect MRAC scheme using a least-squares adaptation method can be found which explicitly minimizes an H_∞ criterion (Hwang and Chen, 1988).

Since the 1960s, MRAC theory has been developed extensively. The application of the stability theories has been extended to the case where only output feedback is used. Several specific discrete-time MRAC algorithms have been developed. Lately, the main research has been focused on weakening the strict assumptions lying at the basis of MRAC design. Since 1980, the robustness of MRAC systems to external and state-dependent disturbances has gained much attention in the literature. Recently, some tools for convergence analysis and order reduction have become available. The main body of MRAC literature is mathematically oriented.

At the same time, many practical applications of MRAC have been reported. It is striking that those algorithms used in practice are all relatively simple, and only a very limited set of theoretical results is used. This is mainly due to the fact that MRAC theory is based on assumptions that are generally not met (for example, the process is required to be linear and of known order). Complex algorithms may achieve improvements in theory, but in practice they make the system more sensitive to violations of the assumptions made. For example, in practice it is observed that an as low as possible number of adjustable parameters is favoured, because convergence is generally slower if more parameters need adjustment. Besides, violation of the mentioned linearity requirement, for example, may prohibit convergence altogether, especially if the number of adjustable parameters is large.

From a theoretical viewpoint, attempts are in progress to bridge the wide gap between theory and practice. This book does not pretend to bring the two worlds together completely. However, in the author's opinion, theoretical developments are only useful if there is a view to practical application. Therefore, this book presents several new strategies which can be applied within existing theories and methods. These strategies are usually based on observations of what can go wrong in practice, trying to find solutions which can be generalized to a class of problems. The strategies found in several cases touch the very heart of MRAC, namely the specification of the desired behaviour by means of a fixed reference model. Several practical applications are presented as illustrations of the developed methods.

1.3 Adaptive controller design

This section introduces some important elements of MRAC systems: the primary controller, the reference model, and the derivation of the adaptive laws. In particular, much attention is paid to the latter, while the adaptive law design takes a central place in the MRAC technique.

1.3.1 The process

Although the process is hardly part of the adaptive controller design, it is the reason for the existence of the adaptive controller-to-be and hence it deserves a place in this section. For design purposes, the process is required to satisfy conditions that it can never meet in real life. For example, to be able to use the adaptive control theory, the process should be linear; the order of its transfer function numerator and denominator polynomials should be known, and the same holds for the sign of the DC gain. In addition, the process transfer function should be minimum phase, and, last but not least, the process should be completely disturbance free.

As mentioned, the process can never satisfy all these requirements. The exact order is never exactly known; there will always be some dynamics that the designer has overlooked while their effect on the process is hardly noticeable. Of course, no process is linear or disturbance free. The sign of the process' DC gain is one of the few exceptions, and generally *is* known. Now, fortunately, the fact that the requirements on the process are never met does not imply that an adaptive controller would not be feasible; it only means that the theory the adaptive controller was designed with was used on a somewhat simplified process model. It is therefore important to know how the adaptive controller operates under nonperfect conditions, i.e. on a real process. This aspect of adaptive control has been the main research subject during the last decade. Chapters 3, 5 and 7 all investigate how an adaptive system reacts to certain types of disturbance, and how the system behaviour can be improved.

1.3.2 The primary controller

The primary controller (see figure 1.2) can in principle have any structure known from linear controller design. There are, however, some important requirements that the primary controller must satisfy.

First, the *perfect model-matching* condition must be satisfied, which states that a controller parameter setting must exist for which the closed-loop behaviour equals the reference model response. This requirement puts demands on the order and structure of the primary controller, in relation to those of the process. For example, a PID algorithm can satisfy the requirement only for processes of low order.

Second, for a direct adaptation to be applicable, the control signal u must be a linear function of the parameters. Normally, the primary controller used in MRAC consists of a signal vector generator part producing a signal vector $\boldsymbol{\omega}$, of which the control u is formed as $u = \boldsymbol{\theta}^T \boldsymbol{\omega}$, where $\boldsymbol{\theta}$ is the controller parameter vector. If the process state vector $\boldsymbol{x}_p$ is completely known, $\boldsymbol{x}_p$ in combination with the reference signal r usually forms the vector $\boldsymbol{\omega}$, and a signal generator part is superfluous. If the process output only can be measured, a signal vector is produced by auxiliary signal generators. The value of the controller parameters $\boldsymbol{\theta}$ for which the closed-loop behaviour matches the reference model is denoted $\boldsymbol{\theta}^*$.

1.3.3 The reference model

The reference model, which specifies the desired process behaviour, is usually given in a parametric form and implemented in a control computer. If state feedback is used, the reference model also produces a complete state vector, whereas in the case of output feedback the desired behaviour is specified as an input–output relationship. The perfect model-matching condition mentioned above places requirements on the reference model, one of which being that its relative degree (the pole excess) must be equal to the relative degree of the process. In addition to the stability, controllability and minimum-phase demands on the reference model, this is the only theoretical requirement. However, it should be noted that in practice, where some theoretical requirements may not be satisfied, the reference model should be chosen sensibly, in the sense that the process output actually can follow the reference model output. If the reference model is chosen to be too fast, for example, the control signal u needs to be extremely large, which causes input saturation effects or unmodelled dynamics to disturb the adaptive system.

Because the actual process capabilities may be unknown or varying, the choice of the reference model is not always obvious. Choosing a very 'cautious' reference model is a solution, but this induces slower closed-loop behaviour than is necessary. The choice of the reference model can be done adaptively by measuring the actual process capabilities and adjusting the reference model accordingly, as is done in the time-optimal control of a direct-drive DC motor discussed in chapter 6. Adjustment of the control goal can also be done directly, for example if unmodelled process dynamics are known to exist, as described in chapter 5.

1.3.4 Derivation of the adaptive laws

A key issue in MRAC design is the derivation of the adaptive laws. The adaptive law derivation can take place using a variety of techniques. This section introduces three important techniques in the more or less chronological order of appearance in the literature. Some general remarks concerning the form of the resulting adaptive laws are given.

The sensitivity approach

The first method that evolved was based on the use of sensitivity models to update the parameters in the correct direction (see, for example, van Amerongen and Honderd, 1983; Narendra and Annaswamy, 1989). As a start, a quadratic performance index C is defined:

$$C(t+T) = \int_t^{t+T} e_1^2(\tau)\,\mathrm{d}\tau$$

In this equation, y_p and y_m are the process and model outputs, respectively, and $e_1 = y_p - y_m$. C is evaluated over a fixed period T, in which the parameters are kept constant. At time $t+T$, a step in the parameters is made in the direction in which C decreases:

$$\begin{aligned}\boldsymbol{\theta}(t+T) &= \boldsymbol{\theta}(t) - \boldsymbol{\Gamma}\frac{\partial C}{\partial \boldsymbol{\theta}} \\ &= \boldsymbol{\theta}(t) - \boldsymbol{\Gamma}\int_t^{t+T} 2e_1(\tau)\frac{\partial e_1(\tau)}{\partial \boldsymbol{\theta}}\,\mathrm{d}\tau \qquad (1.1)\end{aligned}$$

Here, $\boldsymbol{\Gamma}$ must be a square, positive definite matrix, which denotes the adaptation gain. Usually, $\boldsymbol{\Gamma}$ is a diagonal matrix. Making use of:

$$\frac{\partial e_1}{\partial \boldsymbol{\theta}} = \frac{\partial y_p}{\partial \boldsymbol{\theta}}$$

Equation (1.1) becomes:

$$\frac{\boldsymbol{\theta}(t+T) - \boldsymbol{\theta}(t)}{T} = -\frac{1}{T}\boldsymbol{\Gamma}\int_t^{t+T} 2e_1(\tau)\frac{\partial y_p(\tau)}{\partial \boldsymbol{\theta}}\,\mathrm{d}\tau$$

In the limit for $T \to 0$ the adaptive law is obtained:

$$\frac{\mathrm{d}\boldsymbol{\theta}}{\mathrm{d}t} = -2\boldsymbol{\Gamma} e_1 \frac{\partial y_p}{\partial \boldsymbol{\theta}}$$

Note that taking the limit $T \to 0$ involves the use of a criterion function over an infinitely small time interval, which hardly justifies the term 'criterion'. The same adaptive law as above can be obtained by decreasing the momentary quadratic output error $e_1^2(t)$. From an optimization point of view, it is interesting what type of criterion MRAC attempts to minimize. Inclusion of past values of the error in the criterion is not useful because a parameter change at time t cannot alter past error values. This is in contrast with, for example, least-squares process identification, in which a newly found parameter setting also affects past error values. On the other hand, no error prediction is involved in MRAC, and therefore the only possibility left is the use of a 'criterion function' which depends on the error at time t only.

The factor $\partial y_p/\partial \boldsymbol{\theta}$ represents the sensitivity of the process output to variations in $\boldsymbol{\theta}$, and can be generated by a *sensitivity model*. This sensitivity model appears to be equal to a model of the process with the primary controller. Although only the process output is required and thus the method, in principle, would have a wide applicability, several disadvantages can be noted:

- Because the process is not accurately known, a correct sensitivity model cannot be implemented. As a compromise, the reference model parameters are usually used in the sensitivity model, based on the assumption that after some time the system response will approach that of the reference model. This makes the result obtained local in character, and hence the starting values of the primary controller parameters should not differ too much from $\boldsymbol{\theta}^*$.

- The main disadvantage is the absence of a stability proof and, more important, instability can easily occur if, for example, the reference model is not chosen properly or the adaptation gain $\boldsymbol{\Gamma}$ is chosen too large. This disadvantage has led to the use of stability methods in the design of the adaptive laws, as will be described later in this section.

Example 1.1 Derivation of a sensitivity model for a second-order process

Assume a process described by the following input–output relation:

$$\frac{\mathrm{d}^2 y_p}{\mathrm{d}t^2} + a_1 \frac{\mathrm{d}y_p}{\mathrm{d}t} + a_2 y_p = bu$$

In this example, we assume that the process parameters a_1, a_2 and b can be directly adjusted by the adaptation mechanism, i.e. a primary controller structure is not required. This situation is equivalent to the use of complete state feedback, in which parameters of the transfer function can also be influenced directly. The sensitivity coefficients can easily be found by differentiating the process differential equation for the parameters a_1, a_2 and b, respectively. The following three equations then arise:

$$\begin{aligned}
\frac{\mathrm{d}^2}{\mathrm{d}t^2}\left(\frac{\partial y_p}{\partial a_1}\right) + a_1 \frac{\mathrm{d}}{\mathrm{d}t}\left(\frac{\partial y_p}{\partial a_1}\right) + a_2 \frac{\partial y_p}{\partial a_1} &= -\frac{\mathrm{d}y_p}{\mathrm{d}t} \\
\frac{\mathrm{d}^2}{\mathrm{d}t^2}\left(\frac{\partial y_p}{\partial a_2}\right) + a_1 \frac{\mathrm{d}}{\mathrm{d}t}\left(\frac{\partial y_p}{\partial a_2}\right) + a_2 \frac{\partial y_p}{\partial a_2} &= -y_p \\
\frac{\mathrm{d}^2}{\mathrm{d}t^2}\left(\frac{\partial y_p}{\partial b}\right) + a_1 \frac{\mathrm{d}}{\mathrm{d}t}\left(\frac{\partial y_p}{\partial b}\right) + a_2 \frac{\partial y_p}{\partial b} &= u
\end{aligned}$$

These three differential equations have exactly the same coefficients and differ only in their right-hand sides. This result shows that the three sensitivity coefficients $\partial y_p/\partial a_1$, $\partial y_p/\partial a_2$ and $\partial y_p/\partial b$ can be obtained by implementing three identical sensitivity models which have the same coefficients as the process differential equation. Each of these models has a different input, and the output of each model yields one sensitivity coefficient.

It should be noted that in this example the three sensitivity models can be combined into one that is connected to $-y_p$. The output of this sensitivity model yields $\partial y_p/\partial a_2$. The sensitivity coefficient $\partial y_p/\partial a_1$ can be obtained from the internal state of this model. The coefficient $\partial y_p/\partial b$ is equal to y_p/b (compare the sensitivity model for b with the process differential equation: the only difference is the coefficient b in their input), and hence y_p is proportional to this sensitivity coefficient. Assuming that the sign of b is known y_p can be used as the sensitivity coefficient for b.

□

Lyapunov's method

Because of the nonlinear, time-varying character of MRAC systems, linear stability analysis tools cannot be applied. To guarantee the overall stability of the adaptive system, a well-known method is Lyapunov's direct method, which was first used for the purpose of MRAC design by Parks (1966). See also appendix A on stability theory. Lyapunov's method states that a system has a uniform asymptotically stable equilibrium $\boldsymbol{x} = 0$ if a Lyapunov function $V(\boldsymbol{x})$ exists that satisfies:

$$\begin{aligned} V(\boldsymbol{x}) &> 0 \text{ for } \boldsymbol{x} \neq 0 \text{ (positive definite)} \\ \dot{V}(\boldsymbol{x}) &< 0 \text{ for } \boldsymbol{x} \neq 0 \text{ (negative definite)} \\ V(\boldsymbol{x}) &\rightarrow \infty \text{ for } ||\boldsymbol{x}|| \rightarrow \infty \\ V(0) &= 0 \end{aligned} \tag{1.2}$$

Hence, the Lyapunov function V (which has similarities with the *energy content* of the system) must be decreasing with time. Using Lyapunov's method in the adaptive system design, the stability requirements are directly transformed into an adaptive law, in the following way.

- The first step is the derivation of the *error equation*, describing the differential equation of the error signal $\boldsymbol{e}$ (which may be the output error $y_p - y_m$ or the state error $\boldsymbol{x}_p - \boldsymbol{x}_m$). In the derivation of the error equation much freedom exists, but for it to be useable it should be written as a known linear transfer function controlled by a nonlinear input term. Usually, the linear part contains the reference model transfer function. This requirement leads to a more or less standard form of the error equation, as will be described later.

- Second, a Lyapunov function is chosen as a function of both the signal error and the parameter error. This choice is based on a minimal choice of the system states, consisting of the signal error vector $\boldsymbol{e} = \boldsymbol{x}_p - \boldsymbol{x}_m$ and the parameter error vector $\boldsymbol{\phi} = \boldsymbol{\theta} - \boldsymbol{\theta}^*$. In its most simple form, the Lyapunov function V is selected as:

  $$V = \boldsymbol{e}^T \boldsymbol{P} \boldsymbol{e} + \boldsymbol{\phi}^T \boldsymbol{\Gamma}^{-1} \boldsymbol{\phi}$$

  The matrices $\boldsymbol{P}$ and $\boldsymbol{\Gamma}^{-1}$ must be positive definite for V to be an acceptable Lyapunov function candidate.

- Third, the time derivative of the Lyapunov function V is calculated. If $\dot{V}$ is negative definite, uniform asymptotic stability is guaranteed for those variables that $\dot{V}$ is negative definite in. Generally, $\dot{V}$ has the form:

$$\dot{V} = -\boldsymbol{e}^T \boldsymbol{Q} \boldsymbol{e} + \{\text{some terms including } \boldsymbol{\phi}\}$$

By putting the extra terms, including $\boldsymbol{\phi}$, to zero, $\dot{V}$ is guaranteed to be negative definite with respect to $\boldsymbol{e}$ if $\boldsymbol{Q}$ is positive definite. $\boldsymbol{P}$ and $\boldsymbol{Q}$ can be selected with the aid of Lyapunov's theorem, which states that for any asymptotically stable system governed by a matrix $\boldsymbol{A}$, a positive symmetric matrix $\boldsymbol{Q}$ yields a positive symmetric matrix $\boldsymbol{P}$ by the equation:

$$\boldsymbol{A}^T \boldsymbol{P} + \boldsymbol{P}\boldsymbol{A} = -\boldsymbol{Q} \tag{1.3}$$

This equation is referred to as *Lyapunov's equation.* Because the linear part of the error equation usually contains the reference model transfer function, the matrix $\boldsymbol{A}$ in the above formula represents an asymptotically stable system and a positive symmetric matrix $\boldsymbol{Q}$ always yields a positive symmetric matrix $\boldsymbol{P}$.

The negative-definiteness of $\dot{V}$ only occurs as a function of $\boldsymbol{e}$ and not of $\boldsymbol{\phi}$, implying that $\boldsymbol{e} = 0$ is guaranteed to be an asymptotically stable equilibrium point, but $\boldsymbol{\phi} = 0$ is only stable. Hence, convergence of $\boldsymbol{\theta}$ to $\boldsymbol{\theta}^*$ is not guaranteed. However, if the signal vector is 'persistently exciting' (chapter 3), which is the case if the reference signal has a large enough frequency content, $\boldsymbol{\phi} = 0$ is an asymptotically stable equilibrium too. Intuitively this is obvious by considering the following example: if the reference signal is zero, all system states are zero, and so the process states equal the model states. No adaptation takes place and a (incorrect) parameter setting remains. Hence, a certain amount of information must be present to identify all parameters properly.

- Fourth, putting the extra terms to zero directly provides the adaptive laws. Normally, these have the form:

$$\dot{\boldsymbol{\theta}} = -\boldsymbol{\Gamma}\, \epsilon\, \boldsymbol{\xi} \tag{1.4}$$

In equation (1.4), ϵ is directly related to $\boldsymbol{e}$ and $\boldsymbol{\xi}$ is a modified version of $\boldsymbol{\omega}$. These modifications are required to achieve stable adaptive behaviour. ϵ is normally either a linear combination $\boldsymbol{p}^T\boldsymbol{e}$ of the elements of a state error $\boldsymbol{e} = \boldsymbol{x}_p - \boldsymbol{x}_m$, or an augmented version $e_1 + v$ of the output error $e_1 = y_p - y_m$. $\boldsymbol{\xi}$ may then consist of either the signal vector $\boldsymbol{\omega}$ or a filtered version of $\boldsymbol{\omega}$. Chapter 2 will give examples of both possibilities.

Example 1.2 Lyapunov adaptive law design for a first-order system

Consider a first-order proces with unknown, but constant gain b:

$$y_p = \frac{b}{s+1}u$$

Further, we assume a reference model that has the same dynamics as the process:

$$y_m = \frac{1}{s+1}r$$

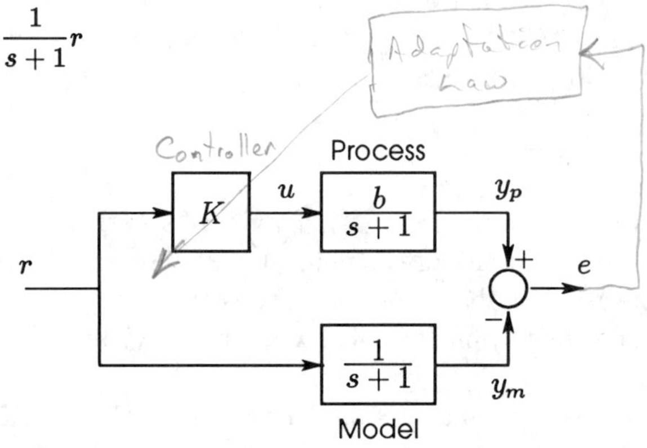

Figure 1.3 Example 1.2, first-order process with reference model

The control u is given by $u = Kr$, in which K is an adjustable gain. A block diagram of this scheme is shown in figure 1.3. The process and model differential equations directly yield the error equation:

$$\begin{aligned} \dot{y}_p &= bKr - y_p \\ \dot{y}_m &= r - y_m \\ \dot{e} = \dot{y}_p - \dot{y}_m &= (bK-1)r - (y_p - y_m) \\ &= -e + (bK-1)r \end{aligned}$$

In this last equation, the term $(bK - 1)$ is proportional to the parameter error $K - 1/b$, and r is the 'signal vector' $\boldsymbol{\omega}$. Now let us choose a Lyapunov function V as:

$$V = e^2 + \frac{1}{\gamma}(bK-1)^2$$

The positive factor γ will appear to be the adaptation gain. Taking the time derivative of V gives:

$$\begin{aligned}\dot{V} &= 2e\dot{e} + \frac{2b}{\gamma}\dot{K}\,(bK-1) \\ &= 2e\left[-e+(bK-1)\,r\right] + \frac{2b}{\gamma}\dot{K}\,(bK-1) \\ &= -2e^2 + 2e\,(bK-1)\,r + \frac{2b}{\gamma}\dot{K}\,(bK-1)\end{aligned}$$

Now the first term, $-e^2$, is negative definite in e. By putting the last two terms to zero, the adaptive law emerges:

$$2e\,(bK-1)\,r + \frac{2b}{\gamma}\dot{K}\,(bK-1) = 0 \implies \dot{K} = -\frac{\gamma}{b}er = -\gamma' er$$

In implementing the adaptive law, γ' must be chosen by the designer. While the adaptive law is stable for all $\gamma > 0$, the sign of b must be known to implement a proper γ'.

□

The main drawback of Lyapunov design is that there is no systematic method of finding a suitable Lyapunov function V which leads to a specific adaptive law. For example, though the designer may intuitively feel that the addition of a proportional factor to the integration will not harm stability, he may have a hard time finding a V which leads to such an adaptive law. The *hyperstability* approach is more flexible in this respect and will be treated next.

Hyperstability

As with Lyapunov's method, an adaptive law designed using hyperstability theory is guaranteed to be stable (Landau, 1979; Parks, 1985). See also appendix A on stability theory. The hyperstability approach is, however, fundamentally different from Lyapunov's method: the designer has to propose an adaptive law, and with the aid of hyperstability theory he can check whether this law gives a stable result.

As in Lyapunov's method, the first step in the design is the derivation of the error equations. These equations are then divided into a linear, time-invariant part and a nonlinear, time-varying part. The first part normally contains the reference model, and its output is the error signal ϵ to be used in the adaptation. The second part contains the adaptive laws and has an output w, which is negated before

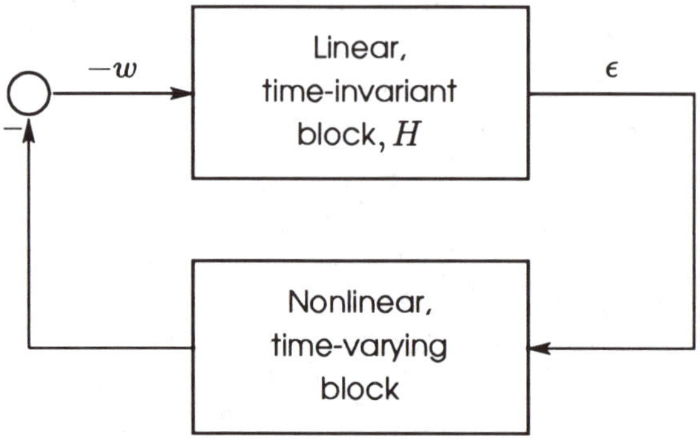

Figure 1.4 Division of error equation into a time-invariant linear part and a time-varying nonlinear part

being fed into the linear part. This division is illustrated in figure 1.4. Usually, the input $-w$ of the linear block equals the multiplication of the parameter error $\boldsymbol{\phi}$ and the signal vector $\boldsymbol{\xi}$ used in the adaptation: $-w = \boldsymbol{\phi}^T\boldsymbol{\xi}$. Hyperstability theory guarantees an asymptotically stable system if both the linear and the nonlinear parts satisfy a passivity condition:

- The linear part H must be strictly positive real (SPR), which means that the real part of $H(j\omega)$ is larger than zero for all $\omega \geq 0$. This implies that the number of poles and zeros in $H(s)$ differs at most by 1, and the phase shift is never larger than 90°. A test for the SPR property of a system in the continuous case can make use of the Kalman–Yakubovich lemma (Slotine and Li, 1991; see also appendix A). In a simplified form, this lemma states that for a strictly positive real transfer function described by matrices $\boldsymbol{A}$, $\boldsymbol{b}$ and $\boldsymbol{c}$, positive definite matrices $\boldsymbol{P}$ and $\boldsymbol{Q}$ can be found that satisfy:

$$\begin{aligned} \boldsymbol{A}^T\boldsymbol{P} + \boldsymbol{P}\boldsymbol{A} &= -\boldsymbol{Q} \\ \boldsymbol{P}\boldsymbol{b} &= \boldsymbol{c} \end{aligned} \tag{1.5}$$

  This lemma is closely related to Lyapunov's equation (1.3), and in the form of equation (1.5) can only be used if the output of the proces to be investigated is not directly dependent on the input (in other words, the $\boldsymbol{d}$ vector in a state-space description is zero). The second equality can be regarded an extension to Lyapunov's equation, and guarantees that the system given by $\boldsymbol{A}$, $\boldsymbol{b}$ and $\boldsymbol{c}$ is not only stable, but also SPR. The above lemma shows that the SPR property of a stable system can be achieved by either modifying $\boldsymbol{b}$, or modifying $\boldsymbol{c}$. In applying Lyapunov's method, the linear part is modified in such a way that it becomes SPR. This is achieved by proper placement of zeros in the linear transfer function by a compensator $\boldsymbol{p}$, as will be clarified in chapter 2.

In discrete time, a simple equivalent for the Kalman–Yakubovich lemma does not exist, and so discrete systems are usually transformed to the s-domain in which the analysis can be accomplished more easily.

- The nonlinear part must satisfy Popov's integral inequality, which states that a positive constant χ exists such that:

$$\int_{\tau=0}^{t} \epsilon(\tau)w(\tau)\,\mathrm{d}\tau \geq -\chi \quad \forall\, t > 0 \tag{1.6}$$

This requirement is also denoted the *passivity* requirement. Observing the nonlinear part as an electrical network, the above inequality can be shown to state that the amount of energy output by the nonlinear system is never larger than the sum of the incoming energy and the energy stored in the system at $t = 0$ (Narendra and Annaswamy, 1989). It is interesting to note that application of Popov's inequality to linear systems leads to the positive-real demand, showing the similarity of the two criteria. A more strict definition gives a condition for *strict passivity*, which in turn is equivalent to strictly positive real transfer functions in the linear case.

Example 1.3 Hyperstability check of example 1.2

In this example we will investigate the hyperstability properties of the adaptive law, derived in example 1.2. The linear part $\dot{e} = -e$ in this case is trivial, while it is a stable first-order system and hence is always SPR.

The nonlinear part produces $w = -\boldsymbol{\phi}^T\boldsymbol{\omega}$ from e. In example 1.2, $\boldsymbol{\phi} = (bK - 1)$ and $\boldsymbol{\omega} = r$ are both scalars. Making use of the adaptive law:

$$K = \int_{\tau=0}^{t} -\frac{\gamma}{b}er\,\mathrm{d}\tau + K(0)$$

Popov's inequality works out as:

$$\begin{aligned}
\int_{t=0}^{T} \epsilon w \, \mathrm{d}t &= -\int_{t=0}^{T} e\,(bK-1)\,r \, \mathrm{d}t \\
&= -\int_{t=0}^{T} er \left[b\left(\int_{\tau=0}^{t} -\frac{\gamma}{b} er \, \mathrm{d}\tau + K(0) \right) - 1 \right] \mathrm{d}t \\
&= \int_{t=0}^{T} er\,\gamma \left[\int_{\tau=0}^{t} er \, \mathrm{d}\tau - \frac{b}{\gamma} K(0) + \frac{1}{\gamma} \right] \mathrm{d}t \\
&= \int_{t=0}^{T} \dot{f}(t) \gamma f(t) \, \mathrm{d}t
\end{aligned}$$

with:

$$f(t) = \int_{\tau=0}^{t} er \, \mathrm{d}\tau - \frac{b}{\gamma} K(0) + \frac{1}{\gamma}$$

Now,

$$\begin{aligned}
\int_{t=0}^{T} \dot{f}(t) \gamma f(t) \, \mathrm{d}t &= \gamma \int_{t=0}^{T} f(t) \, \mathrm{d}f(t) \\
&= \frac{\gamma}{2} \left[f^2(T) - f^2(0) \right] \geq -\frac{\gamma}{2} f^2(0)
\end{aligned}$$

Here, $f(0)$ equals $\left[-\frac{b}{\gamma} K(0) + \frac{1}{\gamma} \right]$. Hence,

$$\int_{t=0}^{T} \epsilon w \, \mathrm{d}t = \gamma \int_{t=0}^{T} f(t) \, \mathrm{d}f(t) \geq -\frac{1}{2\gamma} \left[-bK(0) + 1 \right]^2$$

which proves passivity of the nonlinear part. Note that this reasoning is only valid for a nonzero γ, which implies that an integral part, providing the adaptation 'memory', must be present in the adaptation.

□

Intuitively, the hyperstability concept can best be explained by means of its linear equivalent. Suppose a feedback configuration of a linear feedforward block and a linear feedback block. If both blocks have a phase lag of less than 90°, the loop is stable according to the Nyquist criterion. In fact, one of the blocks is allowed to have a phase lag of 90°, which is important with respect to the adaptation integrators. Any positive loop gain is allowed.

If one of the blocks has more than 90° phase lag, and so the overall phase shift may become larger than 180°, stability is still guaranteed if the loop gain is small. This property also has a counterpart in the nonlinear case, which is called the 'small gain theorem' (Narendra and Annaswamy, 1989; Sastry and Bodson, 1989). This theorem states that if the product of the gains of both blocks is smaller than 1, the loop is always stable. However, the 'gain' in this theorem is defined as the upper bound on the ratio between the output and and input, in which definition a pure integrator has an infinite gain. Because in an MRAC system the nonlinear part usually contains a pure adaptation integrator, the small gain theorem is of limited interest in MRAC. Nevertheless, it can be felt intuitively that if the linear feedforward is not SPR, a small adaptation gain may under some circumstances still provide stable behaviour. This also depends on the violation of the SPR property: if this violation is only marginal, the chance of a stable system is larger. Even if the linear part is not SPR, it is therefore worthwhile to make it as close as possible to being SPR. This observation is important in the case where unmodelled dynamics exist in the process (chapter 5). Section 1.3.5 gives some more insight in this matter by introducing the averaging technique.

In the linear case, one of the blocks is allowed to have more than 90° phase lag, as long as the total phase lag is smaller than 180°. This also has its equivalent in the nonlinear case. A violation of the SPR property is allowed as long as it can be compensated for by a 'superpassive' nonlinear part, an example of which can be found below, in the discussion on proportional adaptation. On the other hand, the nonlinear part need not be passive as long as the linear part is 'superstrictly positive real'. This property is of large interest in discrete adaptive control and allows the use of least-squares adaptation (see chapter 4).

The design based on hyperstability theory can be summarized as follows.

- First, derive the error equations and divide the adaptive system into a linear, time-invariant part (containing the reference model) and a nonlinear part containing the adaptive laws.
- Second, propose an adaptive law which relates the parameter vector $\boldsymbol{\theta}$ to the adaptation error ϵ and the signal vector $\boldsymbol{\xi}$. Check if this adaptive law satisfies Popov's inequality. Normally, the adaptive laws are based on an integrating action, as was found with Lyapunov's method, but they can have various extensions.
- Third, modify the linear part of the error equation such that its transfer function becomes SPR. As in Lyapunov's method, this leads to error compensation or error augmentation, and the Kalman–Yakubovich lemma is a useful tool in this modification.

Normalization

Both the error signal ϵ and the signal vector $\boldsymbol{\xi}$ used in the adaptive law (1.4) are proportionally dependent on the magnitude of the reference signal r, causing a large r to produce faster adaptation than a small r. Therefore, in practice, a normalization is often used, modifying equation (1.4) to:

$$\dot{\boldsymbol{\theta}} = -\boldsymbol{\Gamma} \frac{\epsilon \boldsymbol{\xi}}{1 + \boldsymbol{\xi}^T \boldsymbol{\xi}}$$

The factor '1' in the denominator assures that division by zero can never occur, and in addition has the effect that the adaptation speed is never larger than in the original adaptive law. For small ϵ and $\boldsymbol{\xi}$ the adaptation is not changed, but large ϵ and $\boldsymbol{\xi}$ decrease the adaptation speed. In discrete algorithms, a similar term appears when dealing with the inherent time delay in discrete time, and there it is necessary to guarantee stability.

Other adaptive laws

The integration occurring in the adaptive law (1.4) provides the adaptive system with 'memory', which has the effect that the system 'learns': after some time the error signal ϵ vanishes and the parameter vector $\boldsymbol{\theta}$ approaches a constant value. However, this type of adaptation may have some disadvantages which resemble the disadvantages of a normal, pure, integrating factor in a feedback loop. For example, a limiter in the adaptive loop may induce wind up, a dead zone in the process leads to limit cycles, and although stability is guaranteed, a large $\boldsymbol{\Gamma}$ may lead to (damped) oscillations in the parameter values. Several changes in the above adaptive law can be found in the literature. To increase the convergence of y_p to y_m, a proportional, or even a differentiating term, may be added to the integrator. These extra terms may, however, induce slower convergence of $\boldsymbol{\theta}$ to $\boldsymbol{\theta}^*$. The next paragraph goes into a little more detail regarding proportional adaptation.

Another modification is the replacement of the integrator by logical switches, that switch between two values $\underline{\boldsymbol{\theta}}$ and $\overline{\boldsymbol{\theta}}$. If $\boldsymbol{\theta}^*$ lies somewhere between these values, y_p can be shown to converge to y_m. This modification makes the MRAC scheme resemble a variable-structure design (Itkis, 1976), with accompanying characteristics such as *chatter*. A stability analysis is easily carried out using the hyperstability approach.

Yet another modification, to decrease the negative effects of external disturbances, involves a shift of the adaptation pole to the left half of the s-plane, either permanently or depending on the error magnitude. These types of modification are discussed in detail in chapter 3.

Proportional adaptation

As stated before, in addition to a pure integrating adaptive law (1.4), a proportional term is sometimes used:

$$\boldsymbol{\theta}(t) = \boldsymbol{\theta}(0) - \boldsymbol{\Gamma}_I \int_{\tau=0}^{t} (\epsilon\boldsymbol{\xi})\,\mathrm{d}\tau - \boldsymbol{\Gamma}_P(\epsilon\boldsymbol{\xi})$$

A stability proof of this adaptive law can be obtained by the hyperstability method. The proportional term makes the adaptation react immediately to an error signal and increases the speed with which y_p tends to y_m. However, as stated before, the parameter convergence is generally slower. A proportional term in the adaptation can in some circumstances make the adaptation more stable. For example, in (Tomizuka, 1988) it is shown that, sometimes, a violation of the strictly positive real property of the linear part can be compensated for by proportional adaptation. This, however, only holds for the case where the linear part already has a relative degree of one or zero, which is normally achieved by error compensation or augmentation (see chapter 2). In such a case, the compensation or augmentation can make the linear part SPR as well, limiting the usefulness of Tomizuka's result. The only case in which his result can be usefully applied is when the reference model has a relative degree of one or zero, but is not SPR, which is a rare situation. However, Tomizuka's result illustrates the possibility of exchanging requirements on the linear and the nonlinear parts in the error equation. Note that in the linear equivalent, the addition of a proportional term to the adaptation integrator moves the Nyquist diagram from the imaginary axis to the right half of the complex plane. Hence, the phase shift introduced by the adaptation becomes smaller than 90°, allowing a larger phase shift of the linear part (which is equivalent to no longer requiring a strictly positive real transfer function).

Remarks concerning the adaptive law

Important in equation (1.4) is the multiplication of an error signal by a signal vector, which can be found in all MRAC algorithms. A zero input in the adaptation integrators implies that this product is zero, which leads to the observation that if $\boldsymbol{\theta}$ reaches an equilibrium the *correlation* between the error signal and the signal vector is zero. This means that the parameter setting will approach a value for which the signal vector has no influence on the error signal. If $\boldsymbol{\Gamma}$ is constant, the adaptive law (1.4) is denoted *gradient* adaptation. The adaptation gain matrix $\boldsymbol{\Gamma}$ in continuous MRAC designs is normally a constant positive diagonal matrix; in discrete-time algorithms $\boldsymbol{\Gamma}$ does not always have a diagonal form, and may be recursively adjusted by, for example, a least-squares method (chapter 4).

1.3.5 Averaging techniques

Analyzing properties of MRAC systems is generally not easy, because an MRAC system consists of a set of nonlinear, time-varying differential equations. For investigating stability, Lyapunov's method and the hyperstability approach are useful tools. However, the convergence properties of MRAC, for example, are not easily checked. As an aid in investigating several characteristics of specific MRAC systems, averaging techniques are useful tools (Sastry and Bodson, 1989). Generally speaking, averaging is a way of approximating a set of (possibly nonlinear) time-varying differential equations by a set of (possibly nonlinear) time-invariant equations. This way, given the time-varying system:

$$\dot{\boldsymbol{x}} = \gamma \boldsymbol{f}(t, \boldsymbol{x})$$

the 'averaged' system used for analysis of the original system is given by:

$$\dot{\boldsymbol{x}}_{\mathrm{av}} = \gamma \boldsymbol{f}_{\mathrm{av}}(\boldsymbol{x}_{\mathrm{av}}), \quad \text{with:} \quad \boldsymbol{f}_{\mathrm{av}}(\boldsymbol{x}) = \lim_{T \to \infty} \frac{1}{T} \int_0^T \gamma \boldsymbol{f}(\tau, \boldsymbol{x}) \, \mathrm{d}\tau$$

Hence, the time dependence of the original equation is removed, but a possible nonlinearity in the original equation remains because $\boldsymbol{f}_{\mathrm{av}}$ may still be nonlinear in $\boldsymbol{x}$.

Example 1.4 Averaging of a simple linear system

Consider the following time-varying linear system:

$$\dot{x} = -\gamma \cos^2(t) x$$

The analytic solution of this differential equation is:

$$\begin{aligned} x(t) &= x_0 e^{-\gamma \int_0^t \cos^2(\tau)\, d\tau} \\ &= x_0 e^{-\gamma \int_0^t \left(\frac{1}{2} + \frac{1}{2}\cos 2\tau\right) d\tau} \\ &= x_0 e^{-\frac{1}{2}\gamma t - \frac{\gamma}{4}\sin(2t)} \end{aligned}$$

Here, x_0 is the initial value of $x(t)$ at $t = 0$. The averaged solution is:

$$\dot{x}_{av}(t) = f_{av}(x_{av})$$

with:

$$\begin{aligned} f_{av}(x) &= \lim_{T \to \infty} \frac{1}{T} \int_0^T -\gamma \cos^2(\tau) x \, d\tau \\ &= -\frac{1}{2}\gamma x \end{aligned}$$

and hence $x_{av}(t)$ is:

$$x_{av}(t) = x_0 e^{-\frac{1}{2}\gamma t}$$

The error introduced by the averaging is:

$$x_{av}(t) - x(t) = x_0 e^{-\frac{1}{2}\gamma t} \left[1 - e^{-\frac{\gamma}{4}\sin 2t}\right]$$

This error tends to zero as γ tends to zero, and hence in this simple example where the analytical solution could be calculated, the result obtained with the averaged system approaches that of the real system.

□

In model reference adaptive systems, the equation being averaged is usually the parameter update law, for example:

$$\dot{\boldsymbol{\theta}} = -\boldsymbol{\Gamma}\,\epsilon\,\boldsymbol{\xi} \;\Rightarrow\; \dot{\boldsymbol{\phi}} = -\boldsymbol{\Gamma}\,\epsilon\,\boldsymbol{\xi}$$

In this equation, the time dependence is caused by the signal vector $\boldsymbol{\xi}$, which is a function of the parameter error $\boldsymbol{\phi}(t)$ and the reference $r(t)$. Substituting the equation for the error used in the adaptation: $\epsilon = H(\boldsymbol{\phi}^T\boldsymbol{\xi})$, with H the linear part of the error equation, yields:

$$\dot{\boldsymbol{\phi}} = -\boldsymbol{\Gamma}\boldsymbol{\xi}(\boldsymbol{\phi}, r)H\left[\boldsymbol{\phi}^T\boldsymbol{\xi}(\boldsymbol{\phi}, r)\right] \tag{1.7}$$

The signal vector $\boldsymbol{\xi}$ depends both on the time-varying parameter error $\boldsymbol{\phi}$ and the reference signal r. To be able to apply averaging to the right hand side of equation (1.7), $\boldsymbol{\xi}$ as a function of time should be known. While $\boldsymbol{\xi}$ depends on $\boldsymbol{\phi}(t)$, which is unknown (finding $\boldsymbol{\phi}(t)$ is the aim of the averaging procedure), $\boldsymbol{\phi}$ is assumed to be constant for finding an expression for $\boldsymbol{\xi}$. This assumption is valid if the adaptation acts considerably slower than the signal vector $\boldsymbol{\xi}$. For a constant $\boldsymbol{\phi}$, $\boldsymbol{\xi}(\boldsymbol{\phi}, r)$ can be calculated. Hence, in calculating the quickly varying $\boldsymbol{\xi}$, $\boldsymbol{\phi}$ is assumed to be constant, and hence the resulting fast variations in $\boldsymbol{\xi}$ depend only on the reference $r(t)$ and the process dynamics. These fast variations are averaged:

$$\dot{\boldsymbol{\phi}}_{\text{av}} = -\boldsymbol{\Gamma}\boldsymbol{f}_{\text{av}}\left[\boldsymbol{\phi}_{\text{av}}, \boldsymbol{\xi}(\boldsymbol{\phi}_{\text{av}}, r)\right]$$

Choosing a small adaptation gain $\boldsymbol{\Gamma}$ makes the assumption valid that $\boldsymbol{\phi}$ is fixed.

The averaging procedure can be summarized as follows. The original system consists of a set of differential equations for the signal vector $\boldsymbol{\xi}$ and a set of differential equations for $\boldsymbol{\phi}$. While the differential equation of $\boldsymbol{\xi}$ depends on $\boldsymbol{\phi}$, and the differential equation of $\boldsymbol{\phi}$ depends on $\boldsymbol{\xi}$, both differential equations are time varying.

The signal vector $\boldsymbol{\xi}$ is assumed to vary much faster than the parameter error vector $\boldsymbol{\phi}$. Hence, an expression for $\boldsymbol{\xi}$ can be obtained in which $\boldsymbol{\phi}$ may be considered constant. The time variations in this equation, due to $\boldsymbol{\phi}$, are then removed. As $\boldsymbol{\phi}$ is varying much more slowly than $\boldsymbol{\xi}$, fast variations of $\boldsymbol{\xi}$ are not relevant for the differential equation for $\boldsymbol{\phi}$, and hence the 'averaged' value of $\boldsymbol{\xi}$ may be used in this differential equation.

These considerations make it possible to consider the two differential equations as time invariant. An 'averaged' system emerges which produces an averaged vector ϕ: ϕ_{av}. The averaging technique requires a large separation in time scales between the two sets of differential equations, which is achieved by choosing a small adaptation gain matrix $\boldsymbol{\Gamma}$. The averaged system (which may still be non-linear) is much easier to analyze, for example with respect to the convergence properties.

The main result of the averaging technique is that many important properties of the original system (such as stability and convergence) are directly linked with the properties of the averaged system in the limit for the adaptation gain to zero. For example, if the averaged system is found to be unstable for $\boldsymbol{\Gamma} \rightarrow 0$, the original system is unstable as well. Similarly, the convergence properties of the averaged system correspond with those of the original system. Some important results, obtained by the averaging technique, are discussed below.

Persistently exciting properties

Using the averaging technique, it can easily be derived (Sastry and Bodson, 1989) that the number of spectral lines in the reference signal r must at least be equal to the number of adjustable parameters to achieve convergence. This is a well-known result from identification, but is applicable in adaptive control as well. Further discussion of these persistently exciting properties can be found in chapter 3.

Frequency content in $\boldsymbol{\xi}$

Assume that the ith component of $\boldsymbol{\xi}$ can be written as:

$$\xi_i = \sum_{k=1}^{n_s} a_{ki} \sin(\omega_{ki}t + \phi_{ki}),$$

where n_s is the number of sine functions, a_{ki} the multiplication factor of the kth sine function, and ϕ_{ki} the phase of the kth sine function. Then it can be proven by averaging analysis that the linear part of the error equation H must satisfy (Sastry and Bodson, 1989):

$$\sum_{k=1}^{n_s} \frac{a_{ki}^2}{2} \operatorname{Re}\left[H(j\omega_{ki})\right] > 0 \quad \text{for all} \quad i = 1 \ldots n_p \tag{1.8}$$

in which n_p is the number of elements of $\boldsymbol{\xi}$. The above requirement actually states that for every element of the signal vector, the transfer of the linear part of the error equation must have a net positive real part for the sum of the weights of the frequencies that are present in ξ_i. This result is particularly appealing as it resembles the SPR property on the linear part, but in fact is a less strict condition. If the linear part has a positive real part for the *most important* frequencies in the signal vector, stability is still maintained. This relief of the original SPR condition is due to the summation in equation (1.8), and is based on the assumption of slow adaptation (a small adaptation gain $\boldsymbol{\Gamma}$). Qualitatively, this result is important because it shows that it is worthwhile to increase the frequency area over which H has a positive real part, even if it is impossible to obtain an SPR error equation. This result is of particular interest in chapter 5, where the robustness of MRAC systems in the presence of unmodelled dynamics is under consideration.

1.4 MRAC and other methods

In section 1.2 it was indicated that a definition of MRAC is difficult to give. In the following, some methods are mentioned which are regarded in the literature as being MRAC, and the connection of MRAC with other approaches to controller design is touched upon.

Adaptive H_∞ design

As an alternative to adaptive control, robust controller design is an increasingly important research area. The design of a fixed controller which provides acceptable system behaviour, even in the presence of unmodelled dynamics, external disturbances or unknown parameters, can be based on many approaches. An important class uses minimization of frequency-domain criteria, as in the H_∞ approach. A combination of H_∞ and MRAC was designed by Hwang and Chen (1988), who designed an MRAC system with on-line minimization of an H_∞ criterion of the output error, based on least-squares process identification. The combined analysis of the complete adaptive system allows this scheme to be regarded as an MRAC scheme.

Deadbeat MRAC

Another approach is the steering of the output error to zero in a deadbeat fashion, as presented by Chiang and Chen (1987). A disadvantage of this method is that the reference signal must be known in advance, while it determines the controller parameters.

Optimal control and MRAC

Optimal control strategies, such as LQG, have been linked to MRAC designs by Hwang and Chen (1989). Here, the difference between the reference model and the process is minimized in an LQG sense, which is accomplished by using a least-squares algorithm for process identification and calculating the controller parameters accordingly. This approach has the advantage that the perfect model-matching condition need not be satisfied.

Variable structure systems (VSS)

The VSS approach to controller design bears many resemblances to MRAC, as already mentioned in section 1.3.4 (page 20). In both methods, a desired response is parametrically defined, either by a 'switching surface' or a reference model. In both methods, a multiplication can usually be found of a compensated error signal $\boldsymbol{p}^T\boldsymbol{e}$ and a signal vector. In MRAC, this multiplication is fed into an integrator to calculate the controller parameters, whereas in VSS the multiplication output usually serves as an indication of which control input is to be selected.

The resemblance between VSS and MRAC is stressed even further in a paper by Hsu and Costa (1989), who changed the integrators in a known MRAC scheme to logical switches and obtained a stable control system. The switches can, in fact, be seen as proportional adaptation with an infinite gain over an infinitely small range. Note, however, that although changing integrators to logical switches yields a VSS scheme, VSS methods are not limited to such designs. In fact, the amount of literature on VSS can compete with that on MRAC, and the average reader may not observe any resemblance between the two. In addition, note that the described change to logical switches is only possible if the original MRAC scheme is relatively simple.

Self-tuning controllers

A characteristic of self-tuning controllers is their explicit identification, normally using a least-squares estimation scheme. The aim in this type of system is to minimize a criterion function which includes the difference between the reference generator and the process output, together with a function of the control magnitude. Although the explicit identification can be found only in some discrete-time MRAC algorithms, the *control goal* of self-tuning controllers is obviously related to that of MRAC. A difference is the inclusion of the control magnitude in the criterion function to be minimized, which is normally not present in MRAC schemes. Some special MRAC forms can be shown to be equivalent to existing self-tuning schemes, and some work has been performed on the unification of the two approaches (Johnson, 1980, and Egardt, 1980).

Recent issues in MRAC

In modern literature on MRAC, the main emphasis lies on robustness properties. On the one hand the research focuses on robustness with respect to external disturbances, which can have disastrous effects in MRAC systems. Persistently exciting properties and extensions of stability theories have yielded several interesting theoretical results (chapter 3). On the other hand, the presence of unmodelled dynamics in the process is gaining interest. If the effect of these dynamics is small, simple modifications can cope with them. However, if this is not the case, special precautions have to be taken. Because important dynamics within the bandwidth of the nominal process are usually known to the designer, the problem can in some situations be stated as *model reduction*: the issue is a decrease in complexity of the reference model and the primary controller, while still maintaining acceptable closed-loop behaviour. Other developments include the extension of MRAC systems to the control of nonlinear systems with invertible nonlinearities (for example, see Sastry and Isidori, 1989).

1.5 Outline of the chapters

Although MRAC theory is slowly moving towards a reduction of the strict assumptions lying at the basis of the design, the theoretical results are mostly not in a form that can be applied directly. On the other hand, practical MRAC applications almost always make use of simple controller structures, while a proper operation is achieved by heuristic measures. This book is intended to be part of a link bridging the two viewpoints.

In this book, a relatively large amount of attention is paid to the *reference model*. As stated in section 1.3.4, theoretically the reference model is only required to be of the same relative degree as the process. However, in practice, where the process cannot meet its relatively heavy requirements, the reference model choice is very important. As a simple example, suppose that the reference model is chosen to be so fast that the process input limitations are heavily activated. Because of these input limits, the process can never 'reach' the reference model and a nonzero error remains. However, the parameters are still updated based on this error. This means that an important feedback loop is out of order, namely the influence of a parameter change on the performance (i.e. the error) has vanished. Such a situation may easily lead to parameter drift or, even worse, instability.

An important issue in this book is the observation that, in practice, it is in many cases not sensible to impose a fixed reference model response on to a process, and thus 'lose contact'. For instance, in the above example of input limitation in the process, the reference model is usually chosen such that in the worst case the process can still follow the model. However, in many cases such a choice implies an unnecessarily slow response in situations where the process *can* be driven faster. A better solution might be a continuous adjustment of the reference model to the actual process capabilities. Other examples of this strategy arise when the model cannot be followed exactly by the process, for example in the case of unmodelled process dynamics or nonminimum-phase zeros. In these cases, the reference model can be adjusted for the model-matching capabilities actually present, which may be limited. Adjustment of the reference model to the process (or alternatively modifying the error signal) implies that the reference model is no longer fixed and is only partly chosen by the designer, which seems to conflict with the very basis of MRAC. However, as will become clear in this book, the model adjustment can be regarded as a compromise between what is desired and what is actually possible.

The chapters of this book are organized as follows.

Chapter 2 describes the design of continuous-time MRAC systems. First, the most basic form of MRAC, in which all process state elements are assumed to be accessible, is presented. Because all elements of the state error $\boldsymbol{e} = \boldsymbol{x}_p - \boldsymbol{x}_m$ are then known, application of the stability methods is straightforward. If, however, only the process output is measured, the error equation can no longer be made SPR by a compensation technique. Several methods are mentioned which solve this problem, of which the method using an error-augmenting signal is treated in detail because this is more or less accepted as the standard solution for this type of problem. A modification is presented in the primary controller of this augmented error method, which makes the method resemble other MRAC schemes and is advantageous, especially if some of the process poles are badly damped. Many simulations have led to generally applicable 'tuning rules' for the parameters, to be chosen by the designer. These include, amongst others, the adaptation gains and several filters to be chosen in the adaptive scheme.

Chapter 3 deals with the robustness properties of MRAC systems. In adaptive control, the class of disturbances causing problems is usually different from that in non-adaptive control, leading to a different interpretation of the term 'robustness'. This chapter treats robustness with respect to external disturbances and a few structural disturbances. Some known robustness-improving methods are investigated. The method of adaptive offset compensation is incorporated in the augmented error method, and the new method of using an orthogonal error signal is presented as a way of dealing with dead time in the process transfer function. The definition of the orthogonal error is based on the visual observation that even a small dead time in the process can cause a large output error during transients, while the actual process-model mismatch is small. The orthogonal error is a more realistic measure of the actual difference between process and model, and is the first example of the philosophy this book stands for: *make the error signal insensitive to disturbances that the adaptation should not react to.*

Chapter 4 extends the MRAC design to the discrete-time case. The first discrete-time MRAC algorithms were based on existing continuous-time methods, whereas later developments are completely discrete-time oriented. The modifications of stability theory that least-squares adaptation methods need, in order to guarantee a stable adaptive system, are given. A new method based on a ripple-free primary controller structure is presented which is the second example of the above philosophy. In this case the process zeros are not cancelled in order to avoid possibly harmful effects due to ringing-like problems. The resulting controller appears to have better properties than the original minimum-settling time controller, due especially to a more smoothly starting adaptation. Practical results of application to a water level system complete this chapter.

Chapter 5 introduces the general problem of unmodelled dynamics in the process. An analysis of the effect of these dynamics on the error equation shows where things can go wrong if these unmodelled dynamics persist. For unmodelled dynamics far outside the nominal process bandwidth, some results and modifications are known. However, if the unmodelled dynamics lie within the process bandwidth and have a more than marginal effect on the proces output, the problem setting is slightly different. In this case, the existence of the unmodelled dynamics is usually known to the designer, and hence the term 'unmodelled dynamics' is not completely justified. To overcome terminological difficulties, the term 'structured unmodelled dynamics' is introduced denoting those dynamics in the process transfer that are known to the designer of the adaptive system, but that are disregarded in the design to keep the controller simple.

In order to deal with these 'structured unmodelled dynamics', this chapter presents the method of 'reference model decomposition', which allows inclusion of knowledge of the structure of unmodelled dynamics in the adaptive system. One of the problems induced by unmodelled dynamics in general is a violation of the perfect model-matching requirement, which prohibits the parameters $\boldsymbol{\theta}$ converging to a value for which $y_p = y_m$. By including the knowledge about the unmodelled dynamics in the reference model, and adjusting this part of the model continuously for any mismatch, the negative effects of the unmodelled dynamics on the error equation are relieved. Reference model decomposition allows the designer to use an adaptive controller which is only equipped for the nominal process part, and is therefore of a lower order than would be needed for controlling the full process. The decomposition technique offers a compromise between a not-fulfilled, perfect model-matching condition on the one hand, and instability on the other. Practical results on a gantry crane scale model illustrate the procedure.

Chapter 6 gives a practical application of MRAC to a direct-drive DC motor which shows a similar adjustment of the reference model to the actual process capabilities. In this case, however, the model adjustment is performed adaptively instead of directly. Due to the direct coupling of the load with the motor axis, the desired time-optimal behaviour of the motor depends heavily on the actual load inertia. Because the desired response is not always the same, the reference model cannot be chosen fixed but should adapt itself to the actual motor capabilities. These depend on the load inertia, which is unknown. Therefore, a least-squares algorithm is used to estimate the actual load inertia, which in turn determines the time-optimal behaviour specified in the reference model. This adaptive form of reference model updating complements the direct adjustment presented in chapter 5.

Chapter 7 extends the decomposition approach to the situation where the process suffers from known external disturbances. Here, the situation is considered where two processes with correlated reference signals disturb each other. Because of the correlation between the reference signal and the disturbance, extra cross terms are introduced in the adaptation mechanism. In the case of one-sided interaction, this leads to a decrease in convergence.

Mutual interaction can be considered as time-varying unmodelled dynamics for each of the processes. Because knowledge about the disturbance is assumed to be present, this can be included in the reference model and thus its effect on the output error can be diminished. This leads to improved convergence and may inhibit instability effects. Simulation examples and a practical application to a two-degrees-of-freedom propeller setup play a central role in this chapter.

Chapter 8 reconsiders the aspects treated in this book and tries to give an answer to the questions 'Is MRAC useable in practice?' and 'How about the gap between theory and practice, can it be bridged?'. To stimulate you to read the forthcoming chapters of this book, a hint to the answers is already given: *yes* to both questions.

Problems

1.1 The 'normalization' in the adaptive law, as presented in section 1.3.4, is especially useful in adaptive controller designs which do not allow for a stability proof. Can you explain why?

1.2 In many adaptive schemes, two situations are considered seperately: 1) the process is of first order, and 2) the process is of order two or higher. Can you think of reasons why this is so?

1.3 One of the assumptions in adaptive controller design is that the process parameters are constant or varying slowly. Why is this assumption important for the adaptive controller design?

1.4 In the proposed form of a Lyapunov function V, in section 1.3.4:

$$V = \boldsymbol{e}^T \boldsymbol{P} \boldsymbol{e} + \boldsymbol{\phi}^T \boldsymbol{\Gamma}^{-1} \boldsymbol{\phi}$$

both the signal error $\boldsymbol{e}$ and the parameter error $\boldsymbol{\phi}$ are included. Can you think of reasons why this is necessary?

1.5 As will be treated in more detail in chapter 2, in some MRAC schemes the error $\epsilon = \boldsymbol{p}^T \boldsymbol{e}$ is used in the adaptation, in which $\boldsymbol{p}$ is directly derived from the matrix $\boldsymbol{P}$. Recalling the SPR condition for the linear part of the error equation in the hyperstability approach, can you think of a connection between Lyapunov's method and hyperstability concerning the linear part of the error equation?

1.6 Following the lines of example 1.3, check if the following addition to the integral adaptive law:

$$K = -\gamma' \int_0^t er\, dt - \beta' er, \quad \beta' = \frac{\beta}{b} \text{ with } \beta > 0$$

still conforms to Popov's inequality. Do the same for the addition:

$$K = -\gamma' \int_0^t er\, dt - \rho' \operatorname{sign}(er), \quad \rho' = \frac{\rho}{b} \text{ with } \rho > 0$$

2

Continuous-Time MRAC

MRAC development started with continuous-time analysis methods, and the continuous-time approach is still the basis of most MRAC designs. The main reason for this is that a combined analysis of adaptation and control is not possible for many specifically discrete-time controller structures. Chapter 4 will discuss discrete-time MRAC design.

If the complete process state vector is accessible, the stability theories described in chapter 1 can be applied directly, partly due to the fact that a complete state feedback can be used, making the perfect model-matching condition easily satisfied. If the process output only is available, the design is complicated in two ways. First, the design of a primary controller guaranteeing satisfaction of the perfect model-matching condition is more difficult and, second, to ensure a strictly positive real error equation special measures have to be taken.

2.1 MRAC with complete state information

This section presents an MRAC design for a process of which the complete state vector $\boldsymbol{x}_p$ is available. First, the primary controller structure is given, and following this the adaptive laws are derived.

2.1.1 Primary controller structure

It is assumed that the process is linear and completely controllable, and that it has no zeros. The order of the process is denoted n. A state feedback $u = \boldsymbol{k}_b^T \boldsymbol{x}_p + k_0 r$ allows placement of the closed-loop poles at an arbitrary position, and therefore satisfies the perfect model-matching condition.

The process is described by the state equations:

$$\begin{aligned} \dot{\boldsymbol{x}}_p &= \boldsymbol{A}_p \boldsymbol{x}_p + \boldsymbol{b}_p u \\ y_p &= \boldsymbol{c}^T \boldsymbol{x}_p \end{aligned}$$

$\boldsymbol{A}_p$ is assumed to be in phase-variable form:

$$\begin{aligned} \boldsymbol{A}_p &= \begin{pmatrix} 0 & 1 & 0 & \cdots & 0 \\ 0 & 0 & 1 & \cdots & 0 \\ \vdots & & & \ddots & \vdots \\ -a_{p1} & -a_{p2} & -a_{p3} & \cdots & -a_{pn} \end{pmatrix}, \quad \boldsymbol{b}_p = \begin{pmatrix} 0 \\ \vdots \\ 0 \\ b_{pn} \end{pmatrix}, \\ \boldsymbol{c}^T &= (1,\ 0,\ \ldots,\ 0) \end{aligned} \tag{2.1}$$

Alternatively, the state feedback:

$$u = \boldsymbol{k}_b^T \boldsymbol{x}_p + k_0 r = (k_1,\ k_2,\ \ldots,\ k_n)\, \boldsymbol{x}_p + k_0 r \tag{2.2}$$

can be written as:

$$\begin{aligned} u = \boldsymbol{\theta}^T \boldsymbol{\omega}, \quad \text{with:}\ \boldsymbol{\theta}^T &= \left(k_0,\ \boldsymbol{k}_b^T\right) \\ \text{and:}\ \boldsymbol{\omega}^T &= \left(r,\ \boldsymbol{x}_p^T\right) \end{aligned} \tag{2.3}$$

Equations (2.1) and (2.2) lead to the following description of the closed-loop system:

$$\begin{aligned}
\dot{\boldsymbol{x}}_p &= \left(\boldsymbol{A}_p + \boldsymbol{b}_p \boldsymbol{k}^T\right)\boldsymbol{x}_p + \boldsymbol{b}_p k_0 r \\
&= \begin{pmatrix} 0 & 1 & \cdots & 0 \\ 0 & 0 & \cdots & 0 \\ \vdots & & \ddots & \vdots \\ -a_{p1} + b_{pn}k_1 & -a_{p2} + b_{pn}k_2 & \cdots & -a_{pn} + b_{pn}k_n \end{pmatrix} \boldsymbol{x}_p \\
&\qquad + \begin{pmatrix} 0 \\ \vdots \\ 0 \\ b_{pn}k_0 \end{pmatrix} r \\
&= \boldsymbol{A}_c \boldsymbol{x}_p + \boldsymbol{b}_c r
\end{aligned} \tag{2.4}$$

The process parameters $a_{p1} \ldots a_{pn}$ and b_{pn} are assumed to be unknown but constant. Further, as will be shown in section 2.1.2 (page 40), the sign of b_{pn} must be known. The controller parameters $k_0 \ldots k_n$ can be adjusted by the adaptation mechanism. Note that if the process state vector $\boldsymbol{x}_p$ is not related to a phase-variable form process description, there is a unique transformation which transforms $\boldsymbol{x}_p$ into phase-variable form, and pole placement at an arbitrary location is still possible.

The reference model is described in a similar way to the process:

$$\begin{aligned}
\dot{\boldsymbol{x}}_m &= \boldsymbol{A}_m \boldsymbol{x}_m + \boldsymbol{b}_m r \\
&= \begin{pmatrix} 0 & 1 & 0 & \cdots & 0 \\ 0 & 0 & 1 & \cdots & 0 \\ \vdots & & & \ddots & \vdots \\ -a_{m1} & -a_{m2} & -a_{m3} & \cdots & -a_{mn} \end{pmatrix} \boldsymbol{x}_m + \begin{pmatrix} 0 \\ \vdots \\ 0 \\ b_{mn} \end{pmatrix} r
\end{aligned}$$

Because all elements in the system matrices $\boldsymbol{A}_c$ and $\boldsymbol{b}_c$ that differ from those in the model matrices $\boldsymbol{A}_m$ and $\boldsymbol{b}_m$ can be adjusted, the difference between the closed-loop process transfer function and the model transfer function can be made zero by a proper choice of the controller parameters. The perfect model-matching condition is therefore satisfied. Note that if the process contained zeros, the primary controller structure would have to be extended to be able to cope with these.

Example 2.1 Perfect model matching for a second-order process

In this example, perfect model matching will be considered for a second-order process which is *not* in phase-variable form:

$$\begin{aligned} \dot{\boldsymbol{x}}_p &= \begin{pmatrix} 2 & 4 \\ -4 & -4 \end{pmatrix} \boldsymbol{x}_p + \begin{pmatrix} 0 \\ 1 \end{pmatrix} u \\ y &= \begin{pmatrix} 1 & 0 \end{pmatrix} \boldsymbol{x}_p = x_{p1} \end{aligned}$$

A state feedback is applied:

$$u = k_0 r + k_1 x_{p1} + k_2 x_{p2}$$

which yields the following description of the closed-loop system:

$$\begin{aligned} \dot{\boldsymbol{x}}_p &= \begin{pmatrix} 2 & 4 \\ -4 + k_1 & -4 + k_2 \end{pmatrix} \boldsymbol{x}_p + \begin{pmatrix} 0 \\ k_0 \end{pmatrix} r \\ y &= \begin{pmatrix} 1 & 0 \end{pmatrix} \boldsymbol{x}_p \end{aligned}$$

Now, the question is whether this feedback still makes it possible to reach any second-order reference model. The closed-loop transfer function can be calculated:

$$\begin{aligned} \frac{y}{r} &= \boldsymbol{c}^T (s\boldsymbol{I} - \boldsymbol{A})^{-1} \boldsymbol{b} \\ &= \begin{pmatrix} 1 & 0 \end{pmatrix} \begin{pmatrix} s - 2 & -4 \\ -k_1 + 4 & s - k_2 + 4 \end{pmatrix}^{-1} \begin{pmatrix} 0 \\ k_0 \end{pmatrix} \\ &= \frac{4k_0}{s^2 + (2 - k_2)s + (8 - 4k_1 + 2k_2)} \end{aligned}$$

As can easily be seen, any second-order transfer function can be implemented by a proper choice of the controller parameters and hence the perfect model-matching condition is satisfied. Note, however, that it is not easy to find a reference model of which the internal state can be sensibly compared with x_{p2}. Hence, if possible, the phase-variable form for the process description is preferred in the case of complete state feedback.

□

2.1.2 Derivation of the adaptive laws

The adaptive laws will be derived using Lyapunov's method, following the steps outlined in section 1.3.4.

Derivation of the error equations

For the time derivative of the signal error vector $\boldsymbol{e} = \boldsymbol{x}_p - \boldsymbol{x}_m$ the following equality holds:

$$\begin{aligned}
\dot{\boldsymbol{e}} &= \dot{\boldsymbol{x}}_p - \dot{\boldsymbol{x}}_m \\
&= \boldsymbol{A}_c \boldsymbol{x}_p + \boldsymbol{b}_c r - \boldsymbol{A}_m \boldsymbol{x}_m - \boldsymbol{b}_m r \\
&= \boldsymbol{A}_m (\boldsymbol{x}_p - \boldsymbol{x}_m) + (\boldsymbol{A}_c - \boldsymbol{A}_m)\boldsymbol{x}_p + (\boldsymbol{b}_c - \boldsymbol{b}_m) r \\
&= \boldsymbol{A}_m \boldsymbol{e} + \boldsymbol{A}\boldsymbol{x}_p + \boldsymbol{b} r
\end{aligned} \tag{2.5}$$

with:

$$\begin{aligned}
\boldsymbol{A} &= \boldsymbol{A}_c - \boldsymbol{A}_m \\
\boldsymbol{b} &= \boldsymbol{b}_c - \boldsymbol{b}_m
\end{aligned} \tag{2.6}$$

The parameter error vector $\boldsymbol{\phi}$ is defined as:

$$\boldsymbol{\phi}^T = (b_{pn} k_0 - b_{mn},\ -a_{p1} + b_{pn} k_1 + a_{m1},\ \ldots,\ -a_{pn} + b_{pn} k_n + a_{mn}) \tag{2.7}$$

Using equations (2.7) and (2.3), equation (2.5) can be rewritten as:

$$\begin{aligned}
\dot{\boldsymbol{e}} &= \boldsymbol{A}_m \boldsymbol{e} + \begin{pmatrix} 0 \\ \vdots \\ 0 \\ 1 \end{pmatrix} \boldsymbol{\phi}^T \boldsymbol{\omega} \\
&= \boldsymbol{A}_m \boldsymbol{e} + \boldsymbol{b}_I \boldsymbol{\phi}^T \boldsymbol{\omega}
\end{aligned} \tag{2.8}$$

Equation (2.8) represents the system's error equation, consisting of a linear part governed by $\boldsymbol{A}_m$ and $\boldsymbol{b}_I$, controlled by a nonlinear product term $\boldsymbol{\phi}^T \boldsymbol{\omega}$.

The Lyapunov function

The next step in the design is the choice of the Lyapunov function, and as indicated in chapter 1 this function is normally a quadratic function of both the signal error vector $\boldsymbol{e}$ and the parameter error $\boldsymbol{\phi}$:

$$V = \boldsymbol{e}^T \boldsymbol{P} \boldsymbol{e} + \boldsymbol{\phi}^T \boldsymbol{\Gamma}^{-1} \boldsymbol{\phi} \tag{2.9}$$

The adaptation gain matrix $\boldsymbol{\Gamma}$ must be positive definite and is chosen as a diagonal matrix, so $\boldsymbol{\Gamma}^{-1}$ is positive definite also. $\boldsymbol{P}$ must be a positive definite symmetric matrix and will follow from the adaptive law derivation.

Differentiating V and deriving the adaptive laws

In order to obtain an asymptotically stable adaptive system, $\dot{V}$ must be negative definite. Differentiating V yields:

$$\dot{V} = \boldsymbol{e}^T \left(\boldsymbol{A}_m^T \boldsymbol{P} + \boldsymbol{P} \boldsymbol{A}_m \right) \boldsymbol{e} + 2\boldsymbol{e}^T \boldsymbol{P} \boldsymbol{b}_I \boldsymbol{\phi}^T \boldsymbol{\omega} + 2\boldsymbol{\phi}^T \boldsymbol{\Gamma}^{-1} \dot{\boldsymbol{\phi}} \tag{2.10}$$

By applying Lyapunov's equation (1.3), positive definite symmetric matrices $\boldsymbol{P}$ and $\boldsymbol{Q}$ can be found such that the first part of equation (2.10) satisfies:

$$\boldsymbol{e}^T \left(\boldsymbol{A}_m^T \boldsymbol{P} + \boldsymbol{P} \boldsymbol{A}_m \right) \boldsymbol{e} = -\boldsymbol{e}^T \boldsymbol{Q} \boldsymbol{e}$$

By putting the last two terms of equation (2.10) to zero the adaptive laws emerges:

$$\begin{aligned} 2\boldsymbol{e}^T \boldsymbol{P} \boldsymbol{b}_I \boldsymbol{\phi}^T \boldsymbol{\omega} + 2\boldsymbol{\phi}^T \boldsymbol{\Gamma}^{-1} \dot{\boldsymbol{\phi}} &= 0 \\ \Longrightarrow \dot{\boldsymbol{\phi}} &= -\boldsymbol{\Gamma} \boldsymbol{e}^T \boldsymbol{P} \boldsymbol{b}_I \boldsymbol{\omega} \\ &= -\boldsymbol{\Gamma} \left(\boldsymbol{p}^T \boldsymbol{e} \right) \boldsymbol{\omega} \end{aligned} \tag{2.11}$$

The product $\boldsymbol{P}\boldsymbol{b}_I$ in equation (2.11) is a vector consisting of the nth column $\boldsymbol{p}$ of $\boldsymbol{P}$, and the product of this vector with the signal error vector: $\boldsymbol{p}^T \boldsymbol{e}$ is called the 'compensated error'. This compensated error is used in the adaptive laws to calculate $\dot{\boldsymbol{\phi}}$. The compensator $\boldsymbol{p}$ is regarded as belonging to the linear part of the error equation. In terms of the hyperstability approach, $\boldsymbol{p}$ makes the error equation

SPR, which follows from equation (1.5) by taking $\boldsymbol{b} = \boldsymbol{b}_I$ (the reference model control vector). Making the observation vector $\boldsymbol{c}$ in equation (1.5) equal to the last column of $\boldsymbol{P}$ ensures a strictly positive real linear part of the error equation (see also appendix A).

While the model and process parameters are assumed constant, from the definition of $\boldsymbol{\phi}$ (equation (2.6)) it follows that:

$$
\begin{aligned}
b_{pn}\dot{k}_0 &= -\gamma_{00}\left(\boldsymbol{p}^T\boldsymbol{e}\right) r \\
b_{pn}\dot{k}_i &= -\gamma_{ii}\left(\boldsymbol{p}^T\boldsymbol{e}\right) x_{pi} \quad \text{for} \quad i = 1,\ \ldots,\ n
\end{aligned}
$$

with $\boldsymbol{\Gamma} = \mathrm{diag}(\gamma_{00}, \gamma_{11}, \ldots, \gamma_{nn})$. Then:

$$
\begin{aligned}
\dot{k}_0 &= -\gamma'_{00}\left(\boldsymbol{p}^T\boldsymbol{e}\right) r \\
\dot{k}_i &= -\gamma'_{ii}\left(\boldsymbol{p}^T\boldsymbol{e}\right) x_{pi}
\end{aligned}
\tag{2.12}
$$

or, more generally,

$$
\dot{\boldsymbol{\theta}} = -\boldsymbol{\Gamma}'\left(\boldsymbol{p}^T\boldsymbol{e}\right)\boldsymbol{\omega} \tag{2.13}
$$

with:

$$
\boldsymbol{\Gamma}' = \frac{1}{b_{pn}}\boldsymbol{\Gamma}
$$

The sign of the actual adaptation gain matrix $\boldsymbol{\Gamma}'$ is found to depend on the sign of b_{pn}, and so to be able to implement the adaptive law with a proper sign the sign of b_{pn} must be known. This condition appears in all MRAC schemes. Equations (2.12) form the adaptive laws which provide a stable adaptive system. The matrix $\boldsymbol{P}$, and so the vector $\boldsymbol{p}$, can be calculated with Lyapunov's equation (1.3), starting with a chosen positive definite symmetric matrix $\boldsymbol{Q}$.

Example 2.2 An adaptive system with state feedback

In this example the following process is considered:

$$
\begin{aligned}
\dot{x}_{p1} &= x_{p2} \\
\dot{x}_{p2} &= -4x_{p2} + 2u
\end{aligned}
$$

The reference model is specified as:

$$\begin{aligned}\dot{x}_{m1} &= x_{m2} \\ \dot{x}_{m2} &= -16x_{m1} - 8x_{m2} + 16r\end{aligned}$$

For the primary controller a full state feedback is applied:

$$u = k_0 r + k_1 x_{p1} + k_2 x_{p2}$$

The following matrices apply to the process, the process in closed loop, and the reference model respectively:

$$\begin{aligned}\boldsymbol{A_p} &= \begin{pmatrix} 0 & 1 \\ 0 & -4 \end{pmatrix} & \boldsymbol{b_p} &= \begin{pmatrix} 0 \\ 2 \end{pmatrix} \\ \boldsymbol{A_c} &= \begin{pmatrix} 0 & 1 \\ 2k_1 & 2k_2 - 4 \end{pmatrix} & \boldsymbol{b_c} &= \begin{pmatrix} 0 \\ 2k_0 \end{pmatrix} \\ \boldsymbol{A_m} &= \begin{pmatrix} 0 & 1 \\ -16 & -8 \end{pmatrix} & \boldsymbol{b_m} &= \begin{pmatrix} 0 \\ 16 \end{pmatrix}\end{aligned}$$

Hence, the 'error matrices' $\boldsymbol{A}$ and $\boldsymbol{b}$ are:

$$\begin{aligned}\boldsymbol{A} &= \boldsymbol{A_c} - \boldsymbol{A_m} = \begin{pmatrix} 0 & 0 \\ 2k_1 + 16 & 2k_2 + 4 \end{pmatrix} \\ \boldsymbol{b} &= \boldsymbol{b_c} - \boldsymbol{b_m} = \begin{pmatrix} 0 \\ 2k_0 - 16 \end{pmatrix}\end{aligned}$$

Using the definitions:

$$\begin{aligned}\boldsymbol{\phi}^T &= (2k_0 - 16,\ 2k_1 + 16,\ 2k_2 + 4) \\ \boldsymbol{\omega}^T &= (r,\ x_{p1},\ x_{p2})\end{aligned}$$

the error equation becomes:

$$\dot{\boldsymbol{e}} = \boldsymbol{A_m}\boldsymbol{e} + \begin{pmatrix} 0 \\ \vdots \\ 0 \\ 1 \end{pmatrix} \boldsymbol{\phi}^T \boldsymbol{\omega}$$

Here, $\boldsymbol{e}^T = (e,\ \dot{e})$, with $e = x_{p1} - x_{m1}$ and $\dot{e} = x_{p2} - x_{m2}$. Choosing the standard Lyapunov function (2.9) yields a $\dot{V}$ given by equation (2.10). The first part of $\dot{V}$ yields a positive definite symmetric matrix $\boldsymbol{P}$ by applying Lyapunov's equation, in this example starting with a $\boldsymbol{Q}$ which is equal to the identity matrix:

$$\boldsymbol{A}_m^T \boldsymbol{P} + \boldsymbol{P}\boldsymbol{A}_m = -\boldsymbol{Q}$$

$$\begin{pmatrix} 0 & -16 \\ 1 & -8 \end{pmatrix}\begin{pmatrix} p_{11} & p_{12} \\ p_{12} & p_{22} \end{pmatrix} + \begin{pmatrix} p_{11} & p_{12} \\ p_{12} & p_{22} \end{pmatrix}\begin{pmatrix} 0 & 1 \\ -16 & -8 \end{pmatrix} = \begin{pmatrix} -1 & 0 \\ 0 & -1 \end{pmatrix}$$

$$\begin{pmatrix} -16p_{12} & -16p_{22} \\ p_{11} - 8p_{12} & p_{12} - 8p_{22} \end{pmatrix} + \begin{pmatrix} -16p_{12} & p_{11} - 8p_{12} \\ -16p_{22} & p_{12} - 8p_{22} \end{pmatrix} = \begin{pmatrix} -1 & 0 \\ 0 & -1 \end{pmatrix}$$

$$\begin{pmatrix} -32p_{12} & p_{11} - 8p_{12} - 16p_{22} \\ p_{11} - 8p_{12} - 16p_{22} & 2p_{12} - 16p_{22} \end{pmatrix} = \begin{pmatrix} -1 & 0 \\ 0 & -1 \end{pmatrix}$$

$$\Longrightarrow \begin{cases} p_{11} = \frac{21}{16} \\ p_{12} = \frac{1}{32} \\ p_{22} = \frac{17}{256} \end{cases}$$

Hence,

$$\boldsymbol{P} = \begin{pmatrix} \frac{21}{16} & \frac{1}{32} \\ \frac{1}{32} & \frac{17}{256} \end{pmatrix} \quad \text{or, alternatively,} \quad \boldsymbol{P} = \begin{pmatrix} 336 & 8 \\ 8 & 17 \end{pmatrix}$$

The adaptive laws follow from equation (2.11):

$$\begin{aligned} \dot{\boldsymbol{\phi}} = -\boldsymbol{\Gamma}\boldsymbol{e}^T\boldsymbol{P}\boldsymbol{b}_I\boldsymbol{\omega} &= -\begin{pmatrix} \gamma_0 & 0 & 0 \\ 0 & \gamma_1 & 0 \\ 0 & 0 & \gamma_2 \end{pmatrix}\begin{pmatrix} e & \dot{e} \end{pmatrix}\begin{pmatrix} 8 \\ 17 \end{pmatrix}\begin{pmatrix} r \\ x_{p1} \\ x_{p2} \end{pmatrix} \\ &= -(8e + 17\dot{e})\begin{pmatrix} \gamma_0 r \\ \gamma_1 x_{p1} \\ \gamma_2 x_{p2} \end{pmatrix} \end{aligned}$$

$$\Longrightarrow \quad \begin{cases} \dot{k}_0 = -\gamma_0'(8e + 17\dot{e})\, r \\ \dot{k}_1 = -\gamma_1'(8e + 17\dot{e})\, x_{p1} \\ \dot{k}_2 = -\gamma_2'(8e + 17\dot{e})\, x_{p2} \end{cases}$$

Here, $\gamma_i' = \gamma_i/2$. The factor 2 comes from the process control vector and is generally not known, which does not complicate the implementation while any $\boldsymbol{\Gamma} > 0$ suffices.

□

Example 2.3 Hyperstability check of the method using state feedback

In this example the adaptive laws, derived using Lyapunov's method, are checked for stability by the hyperstability approach, generalizing the result obtained in example 1.3. The linear part of the error equations is:

$$\begin{aligned} \dot{e} &= A_m e + b_I(-w) \\ \epsilon &= p^T e, \quad \text{with } p = P b_I \end{aligned}$$

Because the reference model is in phase-variable form and has a control vector b_I, using p as observation vector directly assures an SPR linear part using the Kalman–Yakubovich lemma (1.5). For investigating the nonlinear part, making use of:

$$\phi = \theta - \theta^* = \int_{t=0}^{T} -\Gamma(p^T e)\omega \, \mathrm{d}t + \theta(0) - \theta^*$$

Popov's inequality becomes:

$$\begin{aligned} \int_{t=0}^{T} \epsilon w \, \mathrm{d}t &= -\int_{t=0}^{T} (p^T e) \left\{ \int_{\tau=0}^{t} -\Gamma(p^T e)\omega \, \mathrm{d}\tau + \theta(0) - \theta^* \right\}^T \omega \, \mathrm{d}t \\ &= \int_{t=0}^{T} \left[(p^T e)\omega\right]^T \Gamma \left\{ \int_{\tau=0}^{t} (p^T e)\omega \, \mathrm{d}\tau - \Gamma^{-1}\left[\theta(0) - \theta^*\right] \right\} \mathrm{d}t \\ &= \int_{t=0}^{T} \dot{f}^T(t) \Gamma f(t) \, \mathrm{d}t \end{aligned}$$

with: $$f(t) = \int_{\tau=0}^{t} (p^T e)\omega \, \mathrm{d}\tau - \Gamma^{-1}\left[\theta(0) - \theta^*\right]$$

Now, by a similar argument as used in example 1.3:

$$\int_{t=0}^{T} \dot{f}^T(t) \Gamma f(t) \, \mathrm{d}t \geq -\frac{1}{2} \Gamma f^T(0) f(0)$$

with:

$$f(0) = -\Gamma^{-1}\left[\theta(0) - \theta^*\right]$$

This shows that both the linear and the nonlinear part of the error equation meet the hyperstability requirements.

□

In the above derivations, and in the examples, it is observed that if the process state vector $\boldsymbol{x}_p$ is available, application of Lyapunov's method leads to an adaptive law in which a linear combination is formed of all elements of the vector $\boldsymbol{e}$. The resulting compensated error is used in the adaptive law. If the process output only is measured, only one element of $\boldsymbol{e}$ is available, and the above procedure cannot be followed. It will appear that in this case the signal vector $\boldsymbol{\omega}$ in equation (2.13) must be altered to maintain a stable adaptive system.

2.2 Extension to the multi-input, multi-output case

The theory developed in the previous section can easily be extended to the multi-input, multi-output case. In fact, while all state elements are assumed to be available, the process outputs are not really important because as $\boldsymbol{x}_p \rightarrow \boldsymbol{x}_m$, $y_p \rightarrow y_m$ automatically. Strictly speaking we are therefore dealing with *multi-input* systems. In this section, we will assume that the process to be controlled consists of two parts with state vectors $\boldsymbol{x}_1$ and $\boldsymbol{x}_2$, respectively, which are mutually dependent. Extension to higher-dimensional MIMO systems is straightforward.

2.2.1 Process and model description

The process is described by the equations:

$$\begin{aligned} \dot{\boldsymbol{x}}_1 &= \boldsymbol{A}_{11}\boldsymbol{x}_1 + \boldsymbol{A}_{12}\boldsymbol{x}_2 + \boldsymbol{b}_{11}u_1 + \boldsymbol{b}_{12}u_2 \\ \dot{\boldsymbol{x}}_2 &= \boldsymbol{A}_{21}\boldsymbol{x}_1 + \boldsymbol{A}_{22}\boldsymbol{x}_2 + \boldsymbol{b}_{21}u_1 + \boldsymbol{b}_{22}u_2 \end{aligned} \tag{2.14}$$

Hence, two inputs u_1 and u_2 act on the process. All matrices $\boldsymbol{A}_{ii}$ are assumed to be in phase-variable form, as in the previous section (equation (2.1)). This implies that the two subsystems each consist of a chain of integrators, the output of each of which influences both the input of the subsystem itself and that of the other subsystem. In other words, each subsystem has the form of the process considered in section 2.1, and is additionally influenced by all states, and the control, of the other subprocess. The separation of the complete process state vector into $\boldsymbol{x}_1$ and $\boldsymbol{x}_2$ seems a bit artificial, but is mainly done because in practical situations such a separation is present. Chapter 7 will assume a similar process structure.

The reference model is assumed to consist of two decoupled parts:

$$\begin{aligned}\dot{\boldsymbol{x}}_{m1} &= \boldsymbol{A}_{m1}\boldsymbol{x}_{m1} + \boldsymbol{b}_{m1}r_1 \\ \dot{\boldsymbol{x}}_{m2} &= \boldsymbol{A}_{m2}\boldsymbol{x}_{m2} + \boldsymbol{b}_{m2}r_2\end{aligned}$$

The reference model matrices $\boldsymbol{A}_{m1}$ and $\boldsymbol{A}_{m2}$ are assumed to be in phase-variable form as well, so that $\boldsymbol{x}_{m1}$ is comparable to $\boldsymbol{x}_1$ and $\boldsymbol{x}_{m2}$ is comparable to $\boldsymbol{x}_2$.

2.2.2 Primary controller structure

The controller structure used is a straightforward extension of the state feedback used in the previous section:

$$\begin{aligned}u_1 &= k_{110}r_1 + k_{120}r_2 + \boldsymbol{k}_{11}^T\boldsymbol{x}_1 + \boldsymbol{k}_{12}^T\boldsymbol{x}_2 \\ u_2 &= k_{210}r_1 + k_{220}r_2 + \boldsymbol{k}_{21}^T\boldsymbol{x}_1 + \boldsymbol{k}_{22}^T\boldsymbol{x}_2\end{aligned} \tag{2.15}$$

Now, extra cross terms k_{120}, k_{210}, $\boldsymbol{k}_{12}$ and $\boldsymbol{k}_{21}$ are introduced to be able to cancel the interaction between the two subsystems. By substituting the expressions for u_1 and u_2 (2.15) in equation (2.14), expressions for the closed loop are found:

$$\begin{aligned}\dot{\boldsymbol{x}}_1 &= \left(\boldsymbol{A}_{11} + \boldsymbol{b}_{11}\boldsymbol{k}_{11}^T + \boldsymbol{b}_{12}\boldsymbol{k}_{21}^T\right)\boldsymbol{x}_1 + \left(\boldsymbol{A}_{12} + \boldsymbol{b}_{11}\boldsymbol{k}_{12}^T + \boldsymbol{b}_{12}\boldsymbol{k}_{22}^T\right)\boldsymbol{x}_2 \\ &\quad + \left(\boldsymbol{b}_{11}k_{110} + \boldsymbol{b}_{12}k_{210}\right)r_1 + \left(\boldsymbol{b}_{11}k_{120} + \boldsymbol{b}_{12}k_{220}\right)r_2 \\ \dot{\boldsymbol{x}}_2 &= \left(\boldsymbol{A}_{21} + \boldsymbol{b}_{22}\boldsymbol{k}_{21}^T + \boldsymbol{b}_{21}\boldsymbol{k}_{11}^T\right)\boldsymbol{x}_1 + \left(\boldsymbol{A}_{22} + \boldsymbol{b}_{22}\boldsymbol{k}_{22}^T + \boldsymbol{b}_{21}\boldsymbol{k}_{12}^T\right)\boldsymbol{x}_2 \\ &\quad + \left(\boldsymbol{b}_{22}k_{210} + \boldsymbol{b}_{21}k_{110}\right)r_1 + \left(\boldsymbol{b}_{22}k_{220} + \boldsymbol{b}_{21}k_{120}\right)r_2\end{aligned}$$

Compactly written, this equates to:

$$\begin{aligned}\dot{\boldsymbol{x}}_1 &= \boldsymbol{A}_{c11}\boldsymbol{x}_1 + \boldsymbol{A}_{c12}\boldsymbol{x}_2 + \boldsymbol{b}_{c11}r_1 + \boldsymbol{b}_{c12}r_2 \\ \dot{\boldsymbol{x}}_2 &= \boldsymbol{A}_{c21}\boldsymbol{x}_1 + \boldsymbol{A}_{c22}\boldsymbol{x}_2 + \boldsymbol{b}_{c21}r_1 + \boldsymbol{b}_{c22}r_2\end{aligned}$$

Here, the subscript 'c' on all matrices and vectors denotes 'closed-loop'. This notation conforms to that in the previous section. Note that all $\boldsymbol{A}_c$ matrices (also the cross matrices $\boldsymbol{A}_{c12}$ and $\boldsymbol{A}_{c21}$) are still in phase-variable form, and all $\boldsymbol{b}_c$ vectors still consist of a series of zeros terminated with one nonzero element, as in the previous section (see equation (2.4)). The parameters of the four control vectors are combined into four vectors $\boldsymbol{\theta}$:

$$\begin{aligned}
\boldsymbol{\theta}_{11}^T &= \left(k_{110},\ \boldsymbol{k}_{11}^T\right) \\
\boldsymbol{\theta}_{12}^T &= \left(k_{120},\ \boldsymbol{k}_{12}^T\right) \\
\boldsymbol{\theta}_{21}^T &= \left(k_{210},\ \boldsymbol{k}_{21}^T\right) \\
\boldsymbol{\theta}_{22}^T &= \left(k_{220},\ \boldsymbol{k}_{22}^T\right)
\end{aligned}$$

In addition, two signal vectors $\boldsymbol{\omega}_1$ and $\boldsymbol{\omega}_2$ are defined:

$$\begin{aligned}
\boldsymbol{\omega}_1^T &= \left(r_1,\ \boldsymbol{x}_1^T\right) \\
\boldsymbol{\omega}_2^T &= \left(r_2,\ \boldsymbol{x}_2^T\right)
\end{aligned}$$

Hence, the two control actions u_1 and u_2 can be rewritten as:

$$\begin{aligned}
u_1 &= \boldsymbol{\theta}_{11}^T \boldsymbol{\omega}_1 + \boldsymbol{\theta}_{12}^T \boldsymbol{\omega}_2 \\
u_2 &= \boldsymbol{\theta}_{21}^T \boldsymbol{\omega}_1 + \boldsymbol{\theta}_{22}^T \boldsymbol{\omega}_2
\end{aligned}$$

2.2.3 Derivation of the adaptive laws

For the derivation of the adaptive laws, two error vectors $\boldsymbol{e}_1$ and $\boldsymbol{e}_2$ are considered:

$$\begin{aligned}
\boldsymbol{e}_1 &= \boldsymbol{x}_1 - \boldsymbol{x}_{m1} \\
\boldsymbol{e}_2 &= \boldsymbol{x}_2 - \boldsymbol{x}_{m2}
\end{aligned}$$

The error equation for $\boldsymbol{e}_1$ is an extended form of the original equation (2.5):

$$\begin{aligned}
\dot{\boldsymbol{e}}_1 &= \boldsymbol{A}_{c11}\boldsymbol{x}_1 + \boldsymbol{A}_{c12}\boldsymbol{x}_2 + \boldsymbol{b}_{c11}r_1 + \boldsymbol{b}_{c12}r_2 - \boldsymbol{A}_{m1}\boldsymbol{x}_{m1} - \boldsymbol{b}_{m1}r_1 \\
&= \boldsymbol{A}_{m1}\left(\boldsymbol{x}_1 - \boldsymbol{x}_{m1}\right) + \left(\boldsymbol{A}_{c11} - \boldsymbol{A}_{m1}\right)\boldsymbol{x}_1 + \left(\boldsymbol{b}_{c11} - \boldsymbol{b}_{m1}\right)r_1 \\
&\quad + \boldsymbol{A}_{c12}\boldsymbol{x}_2 + \boldsymbol{b}_{c12}r_2 \\
&= \boldsymbol{A}_{m1}\boldsymbol{e}_1 + \boldsymbol{A}_1\boldsymbol{x}_1 + \boldsymbol{b}_1 r_1 + \boldsymbol{A}_{c12}\boldsymbol{x}_2 + \boldsymbol{b}_{c12}r_2
\end{aligned}$$

Compared to equation (2.5), this equation has two extra terms $\boldsymbol{A}_{c12}\boldsymbol{x}_2$ and $\boldsymbol{b}_{c12}r_2$, which represent the influence of subsystem 2 on the error of subsystem 1. While the reference model is decoupled, these cross terms are desired to be zero. Fortunately, while $\boldsymbol{A}_{c12}$ and $\boldsymbol{b}_{c12}$ contain the elements of $\boldsymbol{\theta}_{12}$ in the same way as in equation (2.4), these cross terms can be made zero by a proper tuning of $\boldsymbol{\theta}_{12}$. Hence, all nonzero elements of $\boldsymbol{A}_{c12}$ and $\boldsymbol{b}_{c12}$ can be considered part of the parameter errors, as are all nonzero elements of $\boldsymbol{A}_1$ and $\boldsymbol{b}_1$. Accordingly, four parameter error vectors $\boldsymbol{\phi}_{11}$, $\boldsymbol{\phi}_{12}$, $\boldsymbol{\phi}_{21}$ and $\boldsymbol{\phi}_{22}$ can be defined, consisting of the nonzero elements in $(\boldsymbol{A}_1, \boldsymbol{b}_1)$, $(\boldsymbol{A}_{c12}, \boldsymbol{b}_{c12})$, $(\boldsymbol{A}_{c21}, \boldsymbol{b}_{c21})$ and $(\boldsymbol{A}_2, \boldsymbol{b}_2)$, respectively (compare equation (2.7)). These parameter error vectors are directly related to $\boldsymbol{\theta}_{11}$, $\boldsymbol{\theta}_{12}$, $\boldsymbol{\theta}_{21}$ and $\boldsymbol{\theta}_{22}$, in which, however, multiplication factors b_{pn1} and b_{pn2} are present as in the SISO case.

For $\boldsymbol{e}_2$ a similar expression as for $\boldsymbol{e}_1$ is found:

$$\dot{\boldsymbol{e}}_2 = \boldsymbol{A}_{m2}\boldsymbol{e}_2 + \boldsymbol{A}_2\boldsymbol{x}_2 + \boldsymbol{b}_2 r_2 + \boldsymbol{A}_{c21}\boldsymbol{x}_1 + \boldsymbol{b}_{c21} r_1$$

Now, to derive the adaptive laws the following Lyapunov function is chosen:

$$\begin{aligned} V &= \boldsymbol{e}_1^T \boldsymbol{P}_1 \boldsymbol{e}_1 + \boldsymbol{e}_2^T \boldsymbol{P}_2 \boldsymbol{e}_2 \\ &\quad + \boldsymbol{\phi}_{11}^T \boldsymbol{\Gamma}_{11}^{-1} \boldsymbol{\phi}_{11} + \boldsymbol{\phi}_{12}^T \boldsymbol{\Gamma}_{12}^{-1} \boldsymbol{\phi}_{12} + \boldsymbol{\phi}_{21}^T \boldsymbol{\Gamma}_{21}^{-1} \boldsymbol{\phi}_{21} + \boldsymbol{\phi}_{22}^T \boldsymbol{\Gamma}_{22}^{-1} \boldsymbol{\phi}_{22} \end{aligned}$$

This Lyapunov function consists of two parts V_1 and V_2:

$$\begin{aligned} V_1 &= \boldsymbol{e}_1^T \boldsymbol{P}_1 \boldsymbol{e}_1 + \boldsymbol{\phi}_{11}^T \boldsymbol{\Gamma}_{11}^{-1} \boldsymbol{\phi}_{11} + \boldsymbol{\phi}_{12}^T \boldsymbol{\Gamma}_{12}^{-1} \boldsymbol{\phi}_{12} \\ V_2 &= \boldsymbol{e}_2^T \boldsymbol{P}_2 \boldsymbol{e}_2 + \boldsymbol{\phi}_{21}^T \boldsymbol{\Gamma}_{21}^{-1} \boldsymbol{\phi}_{21} + \boldsymbol{\phi}_{22}^T \boldsymbol{\Gamma}_{22}^{-1} \boldsymbol{\phi}_{22} \end{aligned}$$

By showing that each of these functions is negative definite, V is negative definite as well. Taking the derivative of V_1 yields:

$$\begin{aligned} \dot{V}_1 &= \dot{\boldsymbol{e}}_1^T \boldsymbol{P}_1 \boldsymbol{e}_1 + \boldsymbol{e}_1^T \boldsymbol{P}_1 \dot{\boldsymbol{e}}_1 + 2\boldsymbol{\phi}_{11}^T \boldsymbol{\Gamma}_{11}^{-1} \dot{\boldsymbol{\phi}}_{11} + 2\boldsymbol{\phi}_{12}^T \boldsymbol{\Gamma}_{12}^{-1} \dot{\boldsymbol{\phi}}_{12} \\ &= \boldsymbol{e}_1^T \left(\boldsymbol{A}_{m1}^T \boldsymbol{P}_1 + \boldsymbol{P}_1 \boldsymbol{A}_{m1} \right) \boldsymbol{e}_1 \\ &\quad + 2\boldsymbol{e}_1^T \boldsymbol{P}_1 \boldsymbol{A}_1 \boldsymbol{x}_1 + 2\boldsymbol{e}_1^T \boldsymbol{P}_1 \boldsymbol{b}_1 r_1 + 2\boldsymbol{e}_1^T \boldsymbol{P}_1 \boldsymbol{A}_{c12} \boldsymbol{x}_2 + 2\boldsymbol{e}_1^T \boldsymbol{P}_1 \boldsymbol{b}_{c12} r_2 \\ &\quad + 2\boldsymbol{\phi}_{11}^T \boldsymbol{\Gamma}_{11}^{-1} \dot{\boldsymbol{\phi}}_{11} + 2\boldsymbol{\phi}_{12}^T \boldsymbol{\Gamma}_{12}^{-1} \dot{\boldsymbol{\phi}}_{12} \\ &= -\boldsymbol{e}_1^T \boldsymbol{Q}_1 \boldsymbol{e}_1 \\ &\quad + 2\boldsymbol{e}_1^T \boldsymbol{P}_1 \boldsymbol{\phi}_{11}^T \boldsymbol{\omega}_1 + 2\boldsymbol{\phi}_{11}^T \boldsymbol{\Gamma}_{11}^{-1} \dot{\boldsymbol{\phi}}_{11} + 2\boldsymbol{e}_1^T \boldsymbol{P}_1 \boldsymbol{\phi}_{12}^T \boldsymbol{\omega}_2 + 2\boldsymbol{\phi}_{12}^T \boldsymbol{\Gamma}_{12}^{-1} \dot{\boldsymbol{\phi}}_{12} \end{aligned}$$

Putting the last terms to zero, as in the previous section, gives the adaptive laws:

$$\dot{\boldsymbol{\phi}}_{11} = -\boldsymbol{\Gamma}_{11}\left(\boldsymbol{p}_1^T\boldsymbol{e}_1\right)\boldsymbol{\omega}_1$$

$$\dot{\boldsymbol{\phi}}_{12} = -\boldsymbol{\Gamma}_{12}\left(\boldsymbol{p}_1^T\boldsymbol{e}_1\right)\boldsymbol{\omega}_2$$

In exactly the same fashion, adaptive laws for $\boldsymbol{\phi}_{21}$ and $\boldsymbol{\phi}_{22}$ can be obtained:

$$\dot{\boldsymbol{\phi}}_{21} = -\boldsymbol{\Gamma}_{21}\left(\boldsymbol{p}_2^T\boldsymbol{e}_2\right)\boldsymbol{\omega}_1$$

$$\dot{\boldsymbol{\phi}}_{22} = -\boldsymbol{\Gamma}_{22}\left(\boldsymbol{p}_2^T\boldsymbol{e}_2\right)\boldsymbol{\omega}_2$$

Of course, in implementing the adaptive laws for $\boldsymbol{\theta}_{ii}$, the sign of the process coefficients b_{pn1} and b_{pn2} must be known, as in the SISO case. The adaptive laws for the cross term parameter vectors $\boldsymbol{\theta}_{12}$ and $\boldsymbol{\theta}_{21}$ are intuitively easy to interpret. Their input consists of the state vector on which the parameters act, multiplied by an error signal which is actually affected by the adaptation. For example, $\boldsymbol{\theta}_{12}$ is multiplied by $\boldsymbol{\omega}_2$ in the formula for u_1, and hence $\boldsymbol{\omega}_2$ is also present in the adaptive law for $\boldsymbol{\theta}_{12}$. In addition, only $\boldsymbol{e}_1$ is affected by the adaptation of $\boldsymbol{\theta}_{12}$, and hence is present in its adaptive law.

Example 2.4 Application of MIMO MRAC

To illustrate the design of the above-described adaptive control scheme let us apply it to two second-order processes which are mutually coupled:

$$\dot{\boldsymbol{x}}_1 = \begin{pmatrix} 0 & 1 \\ -2 & -8 \end{pmatrix}\boldsymbol{x}_1 + \begin{pmatrix} 0 \\ 8 \end{pmatrix}u_1 + \begin{pmatrix} 0 & 0 \\ 2 & 0 \end{pmatrix}\boldsymbol{x}_2 + \begin{pmatrix} 0 \\ 2 \end{pmatrix}u_2$$

$$\dot{\boldsymbol{x}}_2 = \begin{pmatrix} 0 & 1 \\ -2 & 0 \end{pmatrix}\boldsymbol{x}_2 + \begin{pmatrix} 0 \\ 10 \end{pmatrix}u_2 + \begin{pmatrix} 0 & 0 \\ 1 & 1 \end{pmatrix}\boldsymbol{x}_1 + \begin{pmatrix} 0 \\ 2 \end{pmatrix}u_1$$

In open loop, the combined system is unstable. For both processes, the reference model is chosen as in example 2.2:

$$\dot{\boldsymbol{x}}_{mi} = \begin{pmatrix} 0 & 1 \\ -8 & -16 \end{pmatrix}\boldsymbol{x}_{mi} + \begin{pmatrix} 0 \\ 16 \end{pmatrix}r_i, \quad i = 1,2$$

The primary controller contains twelve parameters:

$$u_1 = k_{110}r_1 + k_{111}x_{11} + k_{112}x_{12} + k_{120}r_2 + k_{121}x_{21} + k_{122}x_{22}$$

$$u_2 = k_{210}r_1 + k_{211}x_{11} + k_{212}x_{12} + k_{220}r_2 + k_{221}x_{21} + k_{222}x_{22}$$

Hence, k_{110}, k_{111}, k_{112}, k_{220}, k_{221} and k_{222} are the state feedback parameters for both subsystems, while k_{120}, k_{121}, k_{122}, k_{210}, k_{211} and k_{212} are decoupling parameters. While the reference model is the same as in example 2.2, the same $\boldsymbol{P}$ matrix can also be used, giving:

$$\boldsymbol{p}_i^T \boldsymbol{e}_i = 8 \cdot e_i + 17 \cdot \dot{e}_i, \quad i = 1, 2$$

Following the derivation as presented above, and using:

$$\begin{aligned} \boldsymbol{k}_{11}^T &= (k_{110},\ k_{111},\ k_{112}) \\ \boldsymbol{k}_{12}^T &= (k_{120},\ k_{121},\ k_{122}) \\ \boldsymbol{k}_{21}^T &= (k_{210},\ k_{211},\ k_{212}) \\ \boldsymbol{k}_{22}^T &= (k_{220},\ k_{221},\ k_{222}) \end{aligned}$$

and:

$$\begin{aligned} \boldsymbol{\omega}_1^T &= (r_1,\ x_{11},\ x_{12}) \\ \boldsymbol{\omega}_2^T &= (r_2,\ x_{21},\ x_{22}) \end{aligned}$$

the adaptive laws become:

$$\begin{aligned} \dot{\boldsymbol{k}}_{11} &= -\boldsymbol{\Gamma}'_{11}(8e_{11} + 17e_{12})\boldsymbol{\omega}_1 \\ \dot{\boldsymbol{k}}_{12} &= -\boldsymbol{\Gamma}'_{12}(8e_{11} + 17e_{12})\boldsymbol{\omega}_2 \\ \dot{\boldsymbol{k}}_{21} &= -\boldsymbol{\Gamma}'_{21}(8e_{21} + 17e_{22})\boldsymbol{\omega}_1 \\ \dot{\boldsymbol{k}}_{22} &= -\boldsymbol{\Gamma}'_{22}(8e_{21} + 17e_{22})\boldsymbol{\omega}_2 \end{aligned}$$

The result is shown in figure 2.1. Block-type setpoints are applied to both loops which have a phase shift to show that the interaction between the loops actually vanishes. All adaptation gains have been set to -0.5. Larger values in this case increase the adaptation speed, but decrease the demonstration effect.

□

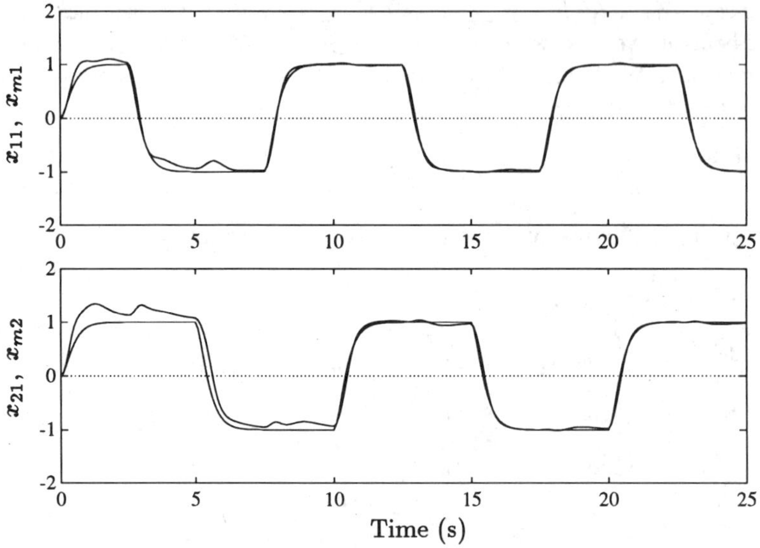

Figure 2.1 Example 2.4, *response obtained using MIMO MRAC to two coupled processes*

2.3 MRAC using output feedback

If the complete process state vector is available, the design procedure for MRAC systems based on stability theory is straightforward, as indicated in the previous section. However, in practice, limited state information may be available, and, for example, only the process output may be measurable. A parallel state reconstruction in such a case, for example by means of a Kalman-filter type of observer, is difficult because of the unknown process parameters.

This section presents several methods available for model reference adaptive control when using output feedback. The following sections treat the 'augmented error method', which is more or less accepted in the literature as the standard approach for MRAC with output feedback.

If only the process output is available, state feedback can no longer be used, and another primary controller must be designed that satisfies the perfect model-matching condition. The derivation of the adaptive laws is also different, because an error compensation as used in the preceding section is not possible due to the absence of the vector $\boldsymbol{e}$. The strictly positive real property of the error equation must therefore be achieved by other means.

An early method that deals with the absence of full state information was presented by Landau (1979). He used state-variable filters to make a reconstruction

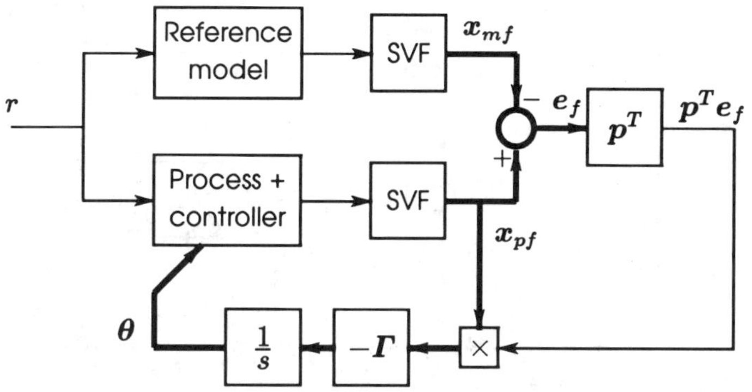

Figure 2.2 Use of state-variable filters by Landau

x_{pf} of the process state x_p, as shown in figure 2.2. To avoid a phase difference between this reconstruction and the model state vector x_m, the same state-variable filters are used for the model output to create x_{mf}. The difference $x_{pf} - x_{mf}$ serves as the error vector e_f. Multiplication of $p^T e_f$ by x_{pf} and use of this signal in the adaptation instead of $(p^T e)x_p$ leads to an asymptotically stable adaptive system. The primary controller structure is, however, not included in Landau's method, and the process parameters are assumed to be able to be adjusted directly. This is an unrealistic assumption because it implies the necessity of complete state information, in which case the method given in section 2.1 can be applied.

Another adaptive scheme was presented by Kaufman and Uliana (1986), who construct the control signal not only of r and y_p, but also include the model state x_m. The error $e_1 = y_p - y_m$ is used in the adaptation (figure 2.3). The primary controller consists of feedforward parameters k_0 and k_m, and one feedback parameter k_e. Due to the inability to influence the error equation such that it becomes SPR, stability can only be assured if the process is SPR too. In that case, however, output feedback by itself would be sufficient to obtain perfect model matching. Although no stability proof can be given for a non-SPR process, in practice the method has an increased stability range because of the inclusion of x_m in u. This inclusion serves as an extra feedforward term for the primary feedback loop, and therefore the loop gain can be kept lower while the process still follows the model. In other words, the feedback gain k_e after convergence will be smaller than it would be if x_m were not included in u. The extra feedforward thus 'steers' the process in the right direction, leading to another parameter set than in the case without such a feedforward.

A third method for MRAC with output feedback uses an augmenting signal

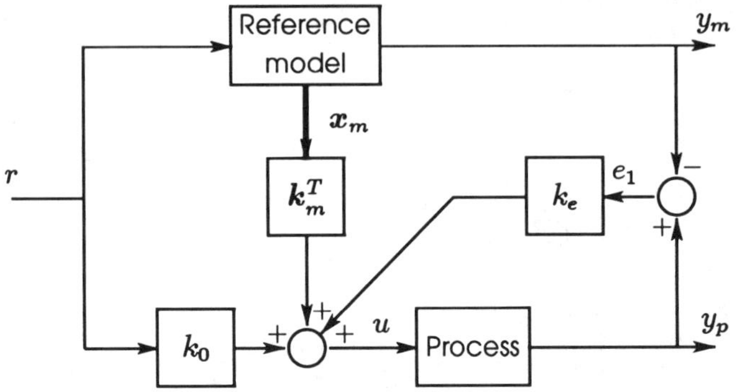

Figure 2.3 Primary controller structure used by Kaufman

v for the output error $e_1 = y_p - y_m$ to compensate for the absence of the complete process state vector $\boldsymbol{x}_p$. This method was first introduced by Monopoli (1974) and is further referred to as the 'augmented error method'. As will be shown, this method has some resemblance to Landau's method in the sense that filters (now called 'auxiliary signal generators') are used to generate signal vectors from y_p and u. However, the way an SPR error equation is achieved differs, and is done here by means of the error augmentation. Later, Narendra and Valavani (1978) presented a slightly different scheme which was essentially equivalent. These approaches did not allow for a proof of the global stability of the adaptive system. The stability problem was solved by Narendra *et al.* (1980); the adaptive laws in these schemes were derived using Lyapunov's method. Suzuki and Dohimoto (1978) showed that by use of hyperstability theory a proportional term can be added to the adaptive laws.

The augmented error method is a combination of a specific primary controller and a set of adaptive laws. In the history of the method, most attention has been paid to the derivation of the latter, while the primary controller structure has remained essentially unchanged. This structure appears to be the cause of slow convergence to the desired response if the process has poles with a low damping ratio. Modifying the signal vector taken from the ASGs will improve the system behaviour in such a case.

2.4 The augmented error method

This section describes the primary controller structure and the adaptive law derivation of the augmented error method.

2.4.1 Primary controller structure

The primary controller structure of the augmented error method is shown in figure 2.4 (Narendra *et al.*, 1980; Narendra and Annaswamy, 1989), which shows that two auxiliary signal generators (ASGs) produce signal vectors $\boldsymbol{\omega}^{(1)}$ and $\boldsymbol{\omega}^{(2)}$. These are both of order $(n-1)$. The signal vectors are multiplied by parameter

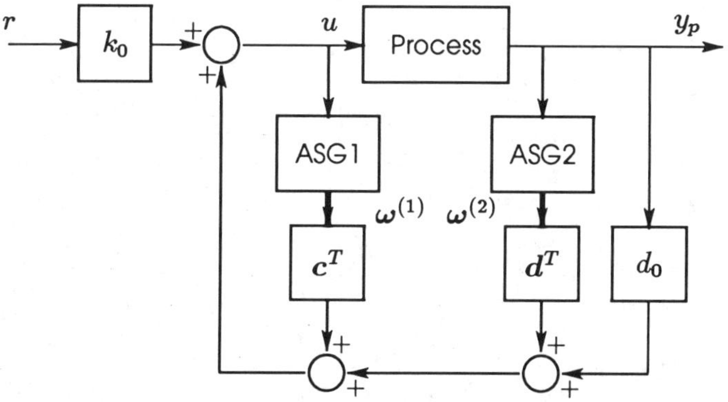

Figure 2.4 Primary controller structure of augmented error method

vectors $\boldsymbol{c}^T$ and $\boldsymbol{d}^T$ and the result is fed back to the process input. The following notation will be used for the process transfer function:

$$W_p = k_p \frac{Z_p}{R_p} \quad Z_p = s^m + \sum_{i=0}^{m-1} b_i s^i$$

$$R_p = s^n + \sum_{i=0}^{n-1} a_i s^i \tag{2.16}$$

Here, $n > m$. The reference model transfer function is:

$$W_m = k_m \frac{Z_m}{R_m}$$

The relative degree of the reference model is assumed to be $n - m$. Further:

$$\boldsymbol{\omega}^T = (r,\ \boldsymbol{\omega}^{(1)T},\ y_p,\ \boldsymbol{\omega}^{(2)T}) \ : \ \text{The signal vector.}$$

$$\boldsymbol{\theta}^T = (k_0,\ \boldsymbol{c}^T,\ d_0,\ \boldsymbol{d}^T) \ : \ \text{The controller parameter vector.}$$

$$\boldsymbol{\theta}^* \ : \ \text{The correct parameter vector.}$$

$$\boldsymbol{\phi} = \boldsymbol{\theta} - \boldsymbol{\theta}^* \ : \ \text{The parameter error vector.}$$

W_p must be minimum phase. In the augmented error method, the auxiliary signal generators take the form of linear filters of order $(n-1)$, as shown in figure 2.5 (the parameters n_i will be explained in section 2.5). This controller structure can be shown to satisfy the perfect model matching condition (see section 2.5).

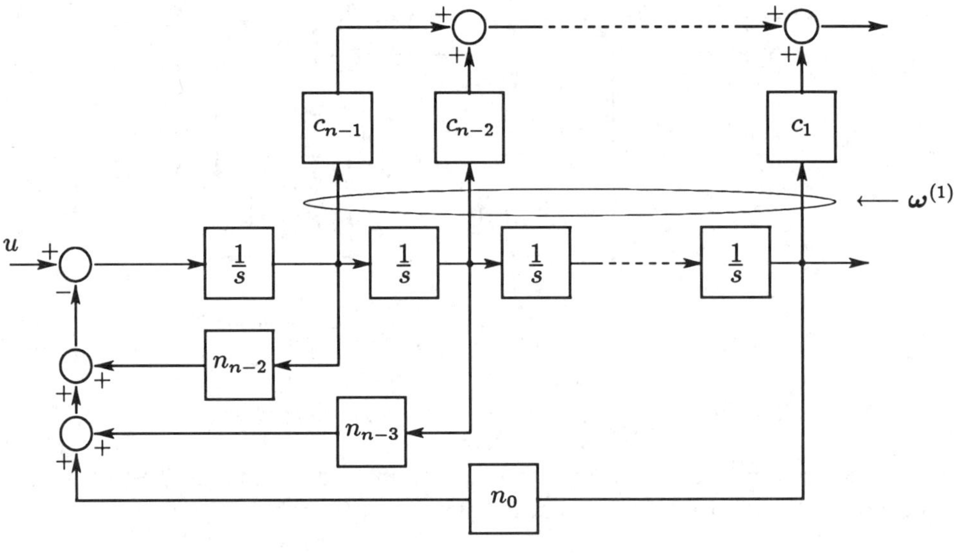

Figure 2.5 Auxiliary signal generator generating $\boldsymbol{\omega}^{(1)}$

2.4.2 Derivation of the adaptive laws

To be able to use either Lyapunov's method or hyperstability theory for the derivation of the adaptive laws, the linear part of the error equation must be strictly positive real. If the error vector $\boldsymbol{e}$ is available, the SPR property is usually obtained by making use of a compensator vector $\boldsymbol{p}$ and using $\epsilon = \boldsymbol{p}^T\boldsymbol{e}$ in the adaptive laws (figure 2.6 (a)). The vector $\boldsymbol{p}$ adds zeros to, or modifies the location of existing zeros of, the error transfer function in such a way that $\epsilon/(-w)$ ($-w$ is the input of the linear part) becomes SPR. However, if only $e_1 = y_p - y_m$ is available, this

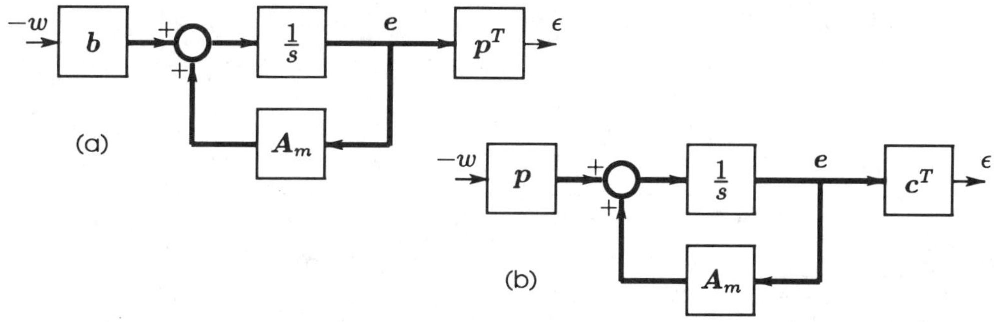

Figure 2.6 Normal place (a) *and alternative place* (b) *of compensator vector* $\boldsymbol{p}$

strategy can no longer be used because then the observation vector $\boldsymbol{c}$ is fixed. To overcome the absence of the complete error vector $\boldsymbol{e}$ it is possible to alter the *input* of the error equations to make the linear part SPR. In other words, the compensator can be placed as an *input vector* of the transfer function, as shown in figure 2.6 (b). This follows directly from the Kalman–Yakubovich lemma (section 1.3.4).

The change of place of the compensator vector $\boldsymbol{p}$ can be accomplished by adding an auxiliary signal to the input of the error model (or, equivalently, to the input of the reference model), as can be clarified as follows. Figure 2.7 shows a general error model including the compensator vector $\boldsymbol{p}$ at the beginning of the transfer function, and an equivalent scheme, indicating that adding an extra term $L(s)(-w)$ to the model input (L containing $p_1 \ldots p_n$) can provide the required SPR property of the transfer function $\epsilon/(-w)$. The augmented error method uses this strategy implicitly, as will be indicated later.

The adaptive laws will now be derived using Lyapunov's second method. It is also possible to apply hyperstability theory to the problem, as described by Suzuki and Dohimoto (1978). The first step in the derivation is the determination of the error equations. This is not as straightforward as in the case of

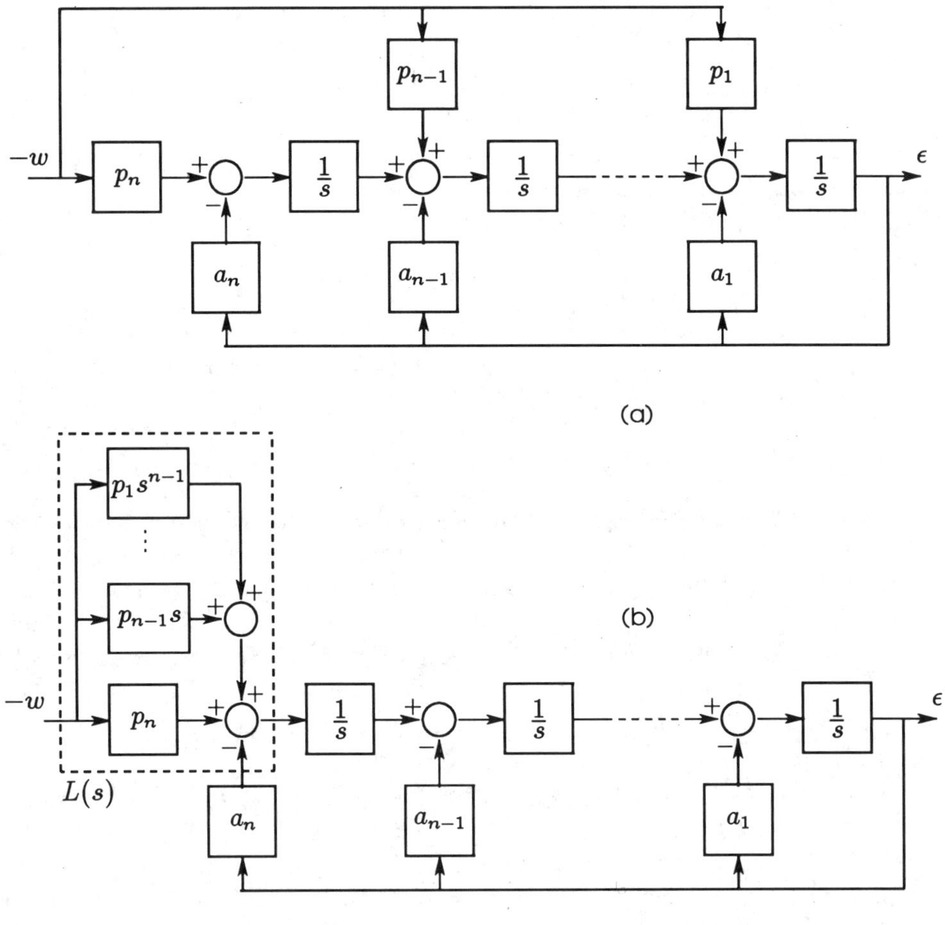

Figure 2.7 General error model (a) *and equivalent scheme* (b)

state feedback because the ASGs introduce extra states in the closed loop. The closed-loop system is then of higher order than the reference model, which complicates the derivation. Narendra and Valavani (1978) derived that the output error $e_1 = y_p - y_m$ can be written as:

$$e_1 = \frac{k_p}{k_m} W_m \left(\boldsymbol{\phi}^T \boldsymbol{\omega}\right) \tag{2.17}$$

In the following, it is assumed that the sign of k_p is known, to be able to impose the SPR property. Without loss of generality (Narendra and Annaswamy, 1989), k_p and k_m are further assumed to be equal. The following equation then holds:

$$y_p = W_m \left(r + \boldsymbol{\phi}^T \boldsymbol{\omega}\right) \tag{2.18}$$

The derivation of the adaptive laws for the cases where W_m is strictly positive real and W_m is not strictly positive real are described separately.

Case 1: **W_m is strictly positive real**

If W_m is strictly positive real no compensator for e_1 is needed and e_1 can be used directly in the adaptive laws. W_m can be written as:

$$W_m = \boldsymbol{c}_I^T \left(s\boldsymbol{I} - \boldsymbol{A}_m\right)^{-1} \boldsymbol{b}_m, \quad \boldsymbol{c}_I^T = (1,\ 0,\ \ldots,\ 0)$$

As in section 1.3.4, a quadratic Lyapunov function is chosen:

$$V = \boldsymbol{e}^T \boldsymbol{P} \boldsymbol{e} + \boldsymbol{\phi}^T \boldsymbol{\Gamma}^{-1} \boldsymbol{\phi}, \quad \boldsymbol{\Gamma}^T = \boldsymbol{\Gamma} > 0 \tag{2.19}$$

In equation (2.19), $\boldsymbol{e}$ is defined as:

$$\boldsymbol{e} = \begin{pmatrix} e_1 \\ \dot{e}_1 \\ \vdots \\ e_1^{(n-1)} \end{pmatrix}$$

Differentiating V yields:

$$\dot{V} = \boldsymbol{e}^T \left(\boldsymbol{A}_m^T \boldsymbol{P} + \boldsymbol{P}\boldsymbol{A}_m\right) \boldsymbol{e} + 2\left(\boldsymbol{\phi}^T \boldsymbol{\omega}\right) \boldsymbol{b}_m^T \boldsymbol{P} \boldsymbol{e} + 2\boldsymbol{\phi}^T \boldsymbol{\Gamma}^{-1} \dot{\boldsymbol{\phi}} \tag{2.20}$$

Because W_m is strictly positive real, according to the Kalman–Yakubovich lemma positive definite symmetric matrices $\boldsymbol{P}$ and $\boldsymbol{Q}$ exist, such that:

$$\begin{aligned} \boldsymbol{A}_m^T\boldsymbol{P} + \boldsymbol{P}\boldsymbol{A}_m &= -\boldsymbol{Q} \\ \boldsymbol{P}\boldsymbol{b}_m &= \boldsymbol{c}_I \end{aligned} \tag{2.21}$$

Now $\boldsymbol{c}_I^T = (1,\ 0,\ \ldots,\ 0)$, and so a positive definite $\boldsymbol{P}$ exists, such that $\boldsymbol{b}_m^T\boldsymbol{P}\boldsymbol{e} = \boldsymbol{c}_I^T\boldsymbol{e} = e_1$. Replacing equation (2.21) in equation (2.20) yields:

$$\dot{V} = -\boldsymbol{e}^T\boldsymbol{Q}\boldsymbol{e} + 2\left(\boldsymbol{\phi}^T\boldsymbol{\omega}\right)e_1 + 2\boldsymbol{\phi}^T\boldsymbol{\Gamma}^{-1}\dot{\boldsymbol{\phi}}$$

To make sure that $\dot{V}$ is (semi) negative definite the last two terms are put to zero, leading to:

$$\left(\boldsymbol{\phi}^T\boldsymbol{\omega}\right)e_1 = -\boldsymbol{\phi}^T\boldsymbol{\Gamma}^{-1}\dot{\boldsymbol{\phi}}$$

The adaptive law then becomes:

$$\dot{\boldsymbol{\theta}} = \dot{\boldsymbol{\phi}} = -\boldsymbol{\Gamma}\boldsymbol{\omega}e_1$$

This is the well-known integral adaptive law (compare equations (1.4) and (2.13)).

Case 2: **W_m is not strictly positive real**

If W_m is not strictly positive real, it is necessary to add zeros to the error equation (2.17). This is achieved by adding an extra signal v (the *auxiliary error*) to the output error e_1:

$$\begin{aligned} \epsilon &= e_1 + v \\ &= e_1 + W_mL\left(\boldsymbol{\phi}L^{-1} - L^{-1}\boldsymbol{\phi}\right)^T\boldsymbol{\omega} \\ &= W_m\left(\boldsymbol{\phi}^T\boldsymbol{\omega}\right) + W_mL\left(\boldsymbol{\phi}L^{-1} - L^{-1}\boldsymbol{\phi}\right)^T\boldsymbol{\omega} \\ &= W_mL\left(\boldsymbol{\phi}^TL^{-1}\boldsymbol{\omega}\right) \end{aligned} \tag{2.22}$$

See notes pg 70 (20 July 93)

The resulting augmented error signal ϵ is used in the adaptation. In the above equation, L is a design polynomial in s, which is chosen such that the product W_mL is SPR. Hence, the order n_L of L must satisfy: $n-m-1 \leq n_L \leq n-m+1$. Another possibility is to choose L a rational function; a popular choice is $L^{-1} = W_m$, leading to an error equation with transfer 1 (see also section 2.8). For the sake of simplicity, in this chapter L will be assumed to be a polynomial of degree $n_L = n - m - 1$.

The adaptive laws are now derived on the basis of the augmented error ϵ instead of the output error e_1. The equation for ϵ includes the polynomial L in the error transfer function. The same Lyapunov function as before is selected:

$$V = \boldsymbol{e}^T \boldsymbol{P} \boldsymbol{e} + \boldsymbol{\phi}^T \boldsymbol{\Gamma}^{-1} \boldsymbol{\phi}$$

In this equation, $\boldsymbol{e}$ is the error vector associated with ϵ instead of e_1:

$$\boldsymbol{e} = \begin{pmatrix} \epsilon \\ \dot{\epsilon} \\ \vdots \\ \epsilon^{(n-1)} \end{pmatrix}$$

The equation for $\boldsymbol{e}$ according to equation (2.22) becomes:

$$\dot{\boldsymbol{e}} = \boldsymbol{A}_m \boldsymbol{e} + \boldsymbol{b}'_m \boldsymbol{\phi}^T (L^{-1} \boldsymbol{\omega}), \qquad \epsilon = (1,\ 0,\ \ldots,\ 0)\, \boldsymbol{e}$$

$b'_m = b_m L$

$\boldsymbol{b}'_m$ differs from $\boldsymbol{b}_m$ in that the zeros introduced by L are included in $\boldsymbol{b}'_m$, because L is part of the new error model (2.22). Differentiating V yields:

$$\dot{V} = \boldsymbol{e}^T \left(\boldsymbol{A}_m^T \boldsymbol{P} + \boldsymbol{P} \boldsymbol{A}_m \right) \boldsymbol{e} + 2 \boldsymbol{b}_m'^T \boldsymbol{P} \boldsymbol{e} \boldsymbol{\phi}^T \left(L^{-1} \boldsymbol{\omega} \right) + 2 \boldsymbol{\phi}^T \boldsymbol{\Gamma}^{-1} \dot{\boldsymbol{\phi}}$$

Because $W_m L$ is SPR, positive definite symmetric matrices $\boldsymbol{P}$ and $\boldsymbol{Q}$ exist that satisfy:

$$\begin{aligned} \boldsymbol{A}_m^T \boldsymbol{P} + \boldsymbol{P} \boldsymbol{A}_m &= -\boldsymbol{Q} \\ \boldsymbol{P} \boldsymbol{b}'_m &= \boldsymbol{c}_I \end{aligned}$$

and so a $\boldsymbol{P}$ exists for which:

$$\boldsymbol{b}_m'^T \boldsymbol{P} \boldsymbol{e} = \boldsymbol{c}_I^T \boldsymbol{e} = \epsilon$$

Requiring that $\dot{V} < 0$ yields:

$$\dot{\boldsymbol{\theta}} = \dot{\boldsymbol{\phi}} = -\boldsymbol{\Gamma} \left(L^{-1} \boldsymbol{\omega} \right) \epsilon$$

Hence it is shown that adding an auxiliary signal as in equation (2.22) allows a stability proof for ϵ and its associated vector $\boldsymbol{e}$. However, the auxiliary term v in equation (2.22) contains the parameter error vector $\boldsymbol{\phi}$, which is, of course, not available. This problem can easily be solved by considering the term:

$$\phi L^{-1} - L^{-1}\phi$$

which is present in equation (2.22). Considering that $\boldsymbol{\phi} = \boldsymbol{\theta} - \boldsymbol{\theta}^*$, this term can be rewritten as:

$$\begin{aligned} \boldsymbol{\phi} L^{-1} - L^{-1}\boldsymbol{\phi} &= (\boldsymbol{\theta} - \boldsymbol{\theta}^*) L^{-1} - L^{-1}(\boldsymbol{\theta} - \boldsymbol{\theta}^*) \\ &= \boldsymbol{\theta} L^{-1} - L^{-1}\boldsymbol{\theta} - \left(\boldsymbol{\theta}^* L^{-1} - L^{-1}\boldsymbol{\theta}^*\right) \end{aligned} \quad (2.23)$$

Because $\boldsymbol{\theta}^*$ is constant, the term between brackets vanishes, and so:

$$\boldsymbol{\phi} L^{-1} - L^{-1}\boldsymbol{\phi} = \boldsymbol{\theta} L^{-1} - L^{-1}\boldsymbol{\theta}$$

Therefore, equation (2.22) can be written as:

$$\epsilon = e_1 + W_m L \left(\boldsymbol{\theta} L^{-1} - L^{-1}\boldsymbol{\theta}\right)^T \boldsymbol{\omega} \quad (2.24)$$

This equation shows that the auxiliary signal can be implemented without differentiators. Note that $u = \boldsymbol{\theta}^T\boldsymbol{\omega}$, and so equation (2.24) can be rewritten as:

$$\epsilon = e_1 + W_m L \left(\boldsymbol{\theta}^T L^{-1}\boldsymbol{\omega} - L^{-1}u\right) \quad (2.25)$$

Figure 2.8 shows a block diagram of the augmented error method with its primary controller above the dashed line and the error-augmenting network below the dashed line. Note that in the adaptive laws, the matrix $\boldsymbol{P}$ is not used, as opposed to the case of state feedback discussed in section 2.1. Only the *existence* of $\boldsymbol{P}$ is needed for the derivation of stable adaptive laws. In the derivation, the zeros introduced by L were considered to exist in the control vector $\boldsymbol{b}'_m$ of the linear part of the error equation. This actually means that L can be considered as the control vector of a state-space description of the error transfer function, as discussed on page 55.

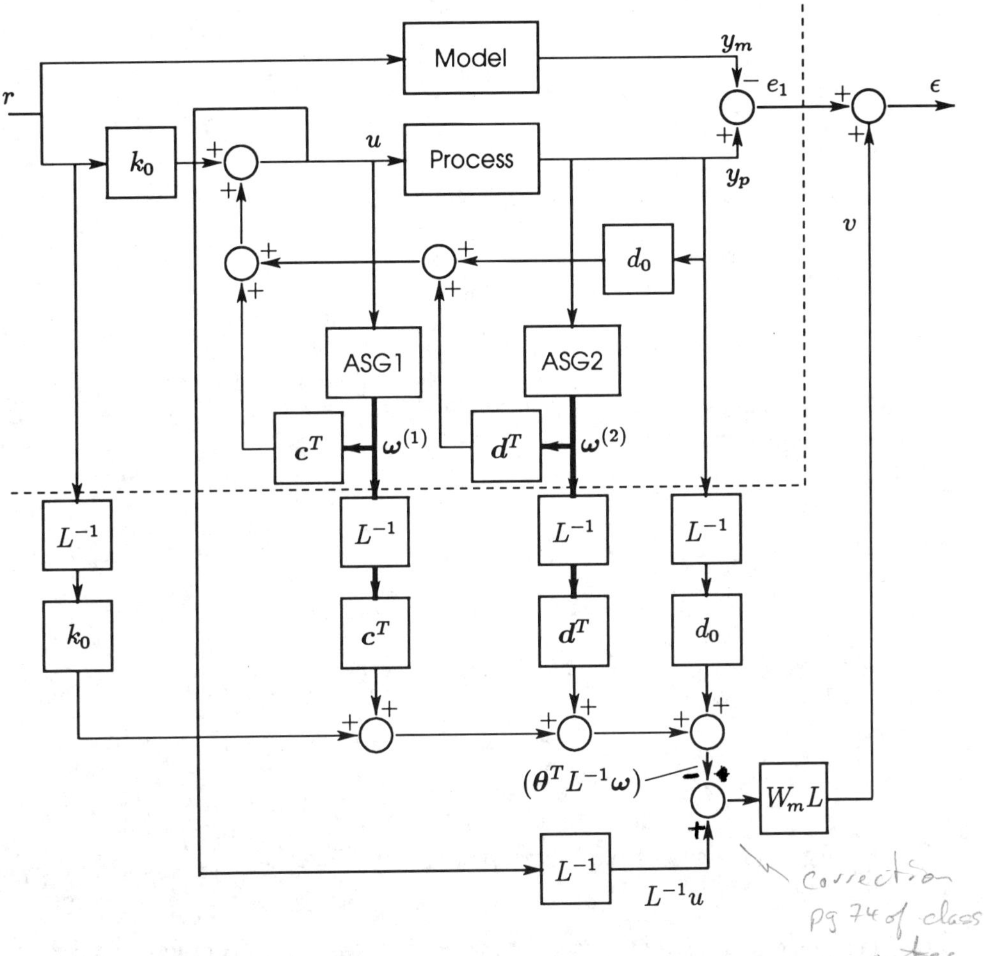

Figure 2.8 General structure of the augmented error method

Example 2.5 Error augmentation for a feedforward parameter

Assume a process which has the same dynamics as the reference model, and an adjustable parameter k_p:

$$y_p = \frac{k_p}{s^2 + 4s + 4} r, \qquad y_m = \frac{4}{s^2 + 4s + 4} r$$

The gain k_p is adjusted in the standard fashion:

$$\dot{k}_p = -\gamma \epsilon r$$

The output error $e = y_p - y_m$ is given by:

$$e = \frac{k_p - 4}{s^2 + 4s + 4} r$$

and hence the linear part of the error equation, $H = 1/(s^2 + 4s + 4)$, is not SPR. To modify e to obtain an ϵ which is connected with an SPR error equation, we apply the error augmentation (2.24), using a filter $L = s + 2$, which makes the linear part of the error equation SPR:

$$\begin{aligned} \epsilon &= e + HL\left(\theta L^{-1} - L^{-1}\theta\right)^T \omega \\ &= e + \frac{1}{s+2}\left(k_p \frac{1}{s+2} r - \frac{1}{s+2} k_p r\right) \end{aligned}$$

The time-varying character of k_p, due to the adaptation, makes the error-augmenting signal nonzero.

□

2.5 An alternative primary controller structure

If some process poles are poorly damped, the above-described augmented error method involves a slow convergence to the desired model response. To discover the cause of this effect, attention must be focused on the auxiliary signal generators as described in section 2.4. As shown there, these generators take the form of linear filters, of which the state is used as the signal vector. However, it seems a more logical approach to use signal generators that give some indication of the actual process state vector x_p, as the state-variable filters in Landau's method do (section 2.3). For example, in a second-order servo system, badly damped

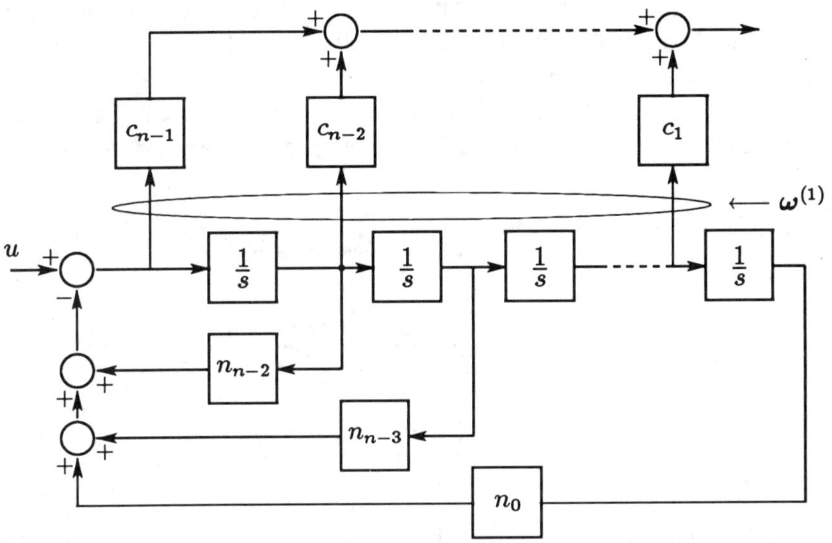

Figure 2.9 Modified use of the auxiliary signal generators

behaviour is usually improved by the addition of feedback of the (estimated) velocity, rather than the filtered position. The modification of the generators is simple, as shown in figure 2.9. Note that all the auxiliary signals generated by the new filter are the derivatives of those generated by the original filter of figure 2.5. The output vector of the modified second auxiliary signal generator consists of $(n-1)$ filtered derivatives of the process output y_p, which can be considered to be an estimation of $\boldsymbol{x}_p$ if the process is described in phase-variable form. In the primary controller, this reconstruction of $\boldsymbol{x}_p$ is fed back to the input of the process, and so resembles the classical use of state feedback. Modification of ASG1, which acts on the process input, has a less obvious effect.

It will now be shown that the proposed change in the auxiliary signal generators is allowed. Because the change only involves a redefinition of the signal vector, it is sufficient to show that the perfect model-matching property of the primary controller is maintained. First, this property will be analyzed for the original controller, and next it will be extended for two modified cases: first, only ASG2 is modified, and then both ASG1 and ASG2 are changed.

Case 1: **Perfect model matching in the original controller**

For the analysis of the perfect model-matching condition in the original case, two methods are available: the analysis can take place by investigating transfer functions and polynomials on the one hand, and by a state-space analysis on the other hand. Although the latter is usually applied in the literature, the first method will be applied here because it allows easier extension to the modified case.

The transfer function of the signal generators, including the parameter vectors $\boldsymbol{c}^T$ and $\boldsymbol{d}^T$ (see figures 2.4 and 2.5), can be written as:

$$
\begin{aligned}
\text{ASG1: } \frac{C}{N}, \text{ with } C &= \underbrace{c_{n-1}s^{n-2} + \ldots + c_2 s + c_1}_{[n-2]} \\
N &= \underbrace{s^{n-1} + \ldots + n_1 s + n_0}_{[n-1]} \\
\text{ASG2: } \frac{D}{N}, \text{ with } D &= \underbrace{d_{n-1}s^{n-2} + \ldots + d_2 s + d_1}_{[n-2]} \qquad (2.26)
\end{aligned}
$$

The coefficients n_i are chosen in such a way that N contains the roots of the reference model numerator polynomial Z_m. The transfer function y_p/r (see figure 2.4) can be written as:

$$
\frac{y_p}{r} = \frac{\overbrace{k_0 k_p Z_p N}^{[n+m-1]}}{\underbrace{(N - C)}_{[n-1]} \underbrace{R_p}_{[n]} - \underbrace{(d_0 N + D)}_{[n-1]} \underbrace{k_p Z_p}_{[m]}} \qquad (2.27)
$$

In this transfer function, the numerator already contains the reference model zeros in N. The process zeros in Z_p must be cancelled by poles, which explains the minimum-phase requirement on W_p. The denominator polynomial is of order $2n - 1$, and it will now be shown that this polynomial is completely determined by the controller parameters, making it possible to use $n - 1$ poles to cancel the process zeros in Z_p and to place n poles at the desired location. To be able to show this, a lemma is needed:

> If Q and T are monic polynomials of degree n_Q and n_T ($\leq n_Q - 1$), respectively, which are relatively prime, polynomials P and R of degree $(n_Q - 1)$ with P monic exist such that $(PQ + RT)$ can be made equal to any arbitrary polynomial of degree $2n_Q - 1$.

A monic polynomial is a polynomial in which the highest coefficient equals one; two polynomials are relatively prime if they have no common roots. To apply this lemma, the denominator of equation (2.27) is compared with the form $PQ + RT$ in the lemma:

$$\begin{aligned} P &= N - C \\ R &= -k_p\,(d_0 N + D) \\ Q &= R_p \\ T &= Z_p \end{aligned}$$

The only requirement on Q and T is that they are both monic, which is the case due to the definition of R_p and Z_p (2.16). It therefore remains to be shown that $(N - C)$ and $[-k_p(d_0 N + D)]$ can be chosen arbitrarily by tuning the controller parameters c_i and d_i. Considering equation (2.26), these polynomials can be written as:

$$\begin{aligned} P &= N - C \\ &= s^{n-1} + (n_{n-2} - c_{n-1})s^{n-2} + \ldots + (n_0 - c_1) \end{aligned} \tag{2.28}$$

$$\begin{aligned} R &= -k_p\,(d_0 N + D) \\ &= -k_p\left[d_0 s^{n-1} + (d_0 n_{n-2} + d_{n-1})\,s^{n-2} + \ldots + (d_0 n_0 + d_1)\right] \end{aligned}$$

N is chosen to contain the reference model zeros and is not influenced by the adaptation, and hence its coefficients n_i $(i = 0 \ldots n-2)$ must be regarded as constant. From equations (2.28) it immediately follows that $(N - C)$ is monic. Its elements can be chosen arbitrarily by selecting c_i $(i = 1 \ldots n-1)$: the ith coefficient is completely determined by c_i. The second polynomial is not monic, but all elements can be selected arbitrarily by the parameters d_i $(i = 0 \ldots n-1)$: d_0 determines the highest ($(n-1)$th) coefficient, and the other parameters determine the remaining coefficients of the polynomial. Both polynomials satisfy the requirements in the above lemma, and hence the denominator polynomial in y_p/r is completely determined by the controller parameters. Note that because the denominator polynomial is monic (while $n > m$), all coefficients are needed to determine the position of the poles, and the gain cannot be chosen arbitrarily. This gain is taken care of by the feedforward parameter k_0 in the numerator of y_p/r.

Case 2: **Perfect model matching if ASG2 is modified**

If ASG2 is modified, D changes to sD due to the redefinition of $\omega^{(2)}$ (like the modification of $\omega^{(1)}$ in figure 2.9). Then the polynomial R changes to:

$$\begin{aligned} R &= -k_p\,(d_0 N + sD) \\ &= -k_p\left[(d_0 + d_{n-1})\,s^{n-1} + (d_0 n_{n-2} + d_{n-2})\,s^{n-2} + \ldots + d_0 n_0\right] \end{aligned}$$

Every element can still be chosen arbitrarily: d_0 determines the lowest coefficient, and d_i $(i = 1 \ldots n-1)$ then determines the remaining coefficients. R therefore still satisfies its requirements and the lemma can be applied as before. The numerator of y_p/r is still uniquely determined by the controller parameters and so the perfect model-matching condition is satisfied.

Case 3: **Perfect model matching if both ASGs are modified**

If both ASGs are modified, C changes to sC in addition to the change of D in case 2, and the polynomial P becomes:

$$\begin{aligned} P &= N - sC \\ &= (1 - c_{n-1})\,s^{n-1} + (n_{n-2} - c_{n-2})s^{n-2} + \ldots + (n_1 - c_1)\,s + n_0 \end{aligned}$$

This polynomial can be rewritten as:

$$P = (1 - c_{n-1})\,P'$$

All coefficients in P' are those of P, divided by $(1 - c_{n-1})$. This makes P' monic and, in addition, allows the selection of every coefficient of P' (except the highest, which is 1 because P' is monic).

The lowest coefficient in P' is determined by c_{n-1} (or, rather, by $n_0/(1 - c_{n-1})$). The polynomial P is now the product of a scalar and a monic polynomial which satisfies the requirements of the lemma presented in case 1. Because R can be chosen arbitrarily, the same holds for $R/(1 - c_{n-1})$. By dividing the resulting denominator polynomial by $(1 - c_{n-1})$, all polynomials in this large polynomial satisfy their requirements, and therefore the denominator can be made equal to any polynomial of degree $2n-1$, except for a factor $(1-c_{n-1})$. Because the parameter k_0 is present in the numerator of y_p/r, k_0 can compensate for the factor $(1-c_{n-1})$, and therefore the perfect model matching condition is still satisfied. Note that c_{n-1} may not have a value of 1.

Although modification of both ASG1 and ASG2 is allowed, only modification of ASG2 is intuitively appealing because it resembles the principle of state feedback. Modification of ASG1 has no such effect. Besides, the implementation of a modified ASG1 causes problems due to the occurrence of an algebraic loop in the calculation of the control u, and the fact that c_{n-1} may not become 1. Simulations have shown that modification of ASG1 hardly influences the system behaviour. For these reasons, this modification is omitted in all experiments.

Example 2.6 Original and modified use for second-order process

A simple example of the original and modified use of the auxiliary signal generators is shown in figure 2.10. Originally, the output y_p of a second-order process and its filtered value y_{pf} are used as feedback signals (a). The filtered process output y_{pf} closely resembles y_p if the filtering constant α is large, and in this case feedback of y_{pf} is not useful. If α is small, y_{pf} has a considerable phase lag compared to y_p, which is not desirable in a feedback loop. Figure 2.10 also

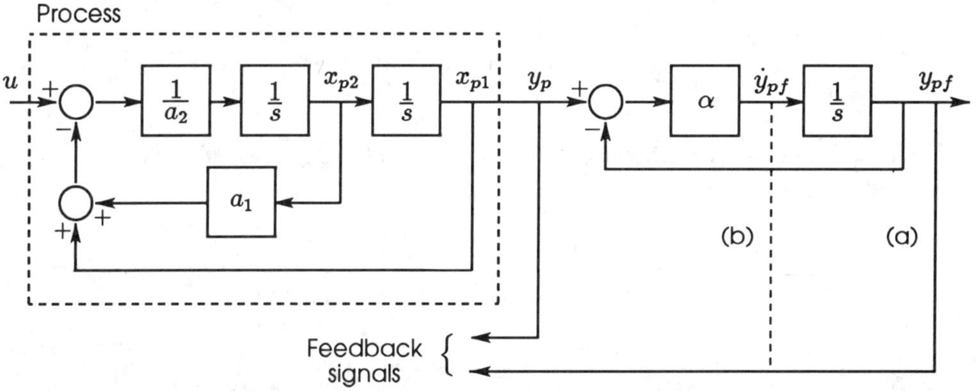

Figure 2.10 Example 2.6, *original* (a) *and modified use* (b) *of the second auxiliary signal generator, using a second-order process*

shows the modified use of the auxiliary signal generators (the dotted line (b)). In this structure, y_p and its filtered derivative $\dot{y}_{pf}$ are fed back to the input of the process. The value of $\dot{y}_{pf}$ can be considered an estimate of the internal second process state x_{p2}. This modified structure is similar to state feedback, in which the poles of the system can be placed at an arbitrary location. Note that $\dot{y}_{pf}$ need not necessarily be a *good* estimation of x_{p2}. Perfect model matching can be achieved with any α.

□

The described change appears to be most valuable if the process has poles with a low damping ratio. In such a situation, feedback of the extra estimated process state can provide the necessary stability. In other cases, the results with and without ASG modification are comparable.

2.6 Simulation examples

This section presents some simulation results obtained using the augmented error method. The first simulations were performed using a second-order process with transfer function:

$$W_p = \frac{12}{s^2 + 2s + 8} \qquad \text{with poles at: } s = -1 \pm 2.65j$$

And a reference model:

$$W_m = \frac{16}{s^2 + 8s + 16} \qquad \text{with poles at: } s = -4 \ \ (2\times)$$

Figure 2.11 shows the response of the model and the process to a block input without adaptation.

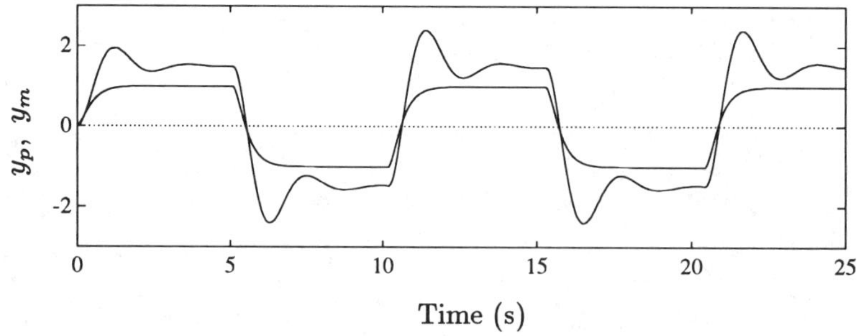

Figure 2.11 Responses of model and process to a block-type reference signal, obtained without adaptation

The auxiliary signal generators are chosen as in figure 2.10. The parameter α in these generators can theoretically be chosen arbitrarily, because the reference model has no zeros. Making α large seems not to be useful, because y_{pf} then closely resembles y_p. In practice, the choice of α has hardly any effect on the response. If the auxiliary signal generators are modified as described in section 2.5, making α large improves the accuracy with which the second process state is estimated and is thus expected to improve the system's response. To avoid disturbances in the estimation due to measurement noise (see chapter 3), α should be chosen such that the bandwidth of the auxiliary signal generators is not considerably larger than that of the process. During the simulations, in which no measurement noise is present, α has a value of 10.

For the filters L^{-1}, a first-order transfer function $1/(s+l)$ is used. The parameter l must be chosen carefully in view of the stability proof, which demands that $W_m L$ is SPR. The reference model has two poles at -4, and an analysis of the phase ϕ of the system:

$$H = \frac{s+l}{(s+4)^2}, \quad \phi = \arctan(\omega/l) - 2\arctan(\omega/4)$$

shows that the added zero may lie between $s = 0$ and $s = -8$. The simulations have been performed with $l = 4$. Figures 2.12 and 2.13 show the response using the original and modified augmented error method respectively. The original response has been obtained by using the following adaptation gains:

$$\begin{aligned} \gamma_{k_0} &= -0.25 \\ \gamma_{c_1} &= -0.125 \\ \gamma_{d_0} &= -0.25 \\ \gamma_{d_1} &= -0.5 \end{aligned}$$

Increasing the adaptation gains does not improve the convergence of the system. In using the modified ASG2, the adaptation gains were chosen a factor 8 larger, and in that case a larger gain gave a significant improvement.

To verify the obtained results further, both the original and the modified augmented error methods were applied to a third-order process:

$$W_p = \frac{20.25}{s^3 + 4s^2 + 7.5s + 13.5} \qquad \text{with poles at: } \begin{aligned} s &= -3 \\ s &= -0.5 \pm 2j \end{aligned}$$

The reference model has a transfer function:

$$W_m = \frac{128}{s^3 + 16s^2 + 80s + 128} \qquad \text{with poles at: } \begin{aligned} s &= -4 \ (2\times) \\ s &= -8 \end{aligned}$$

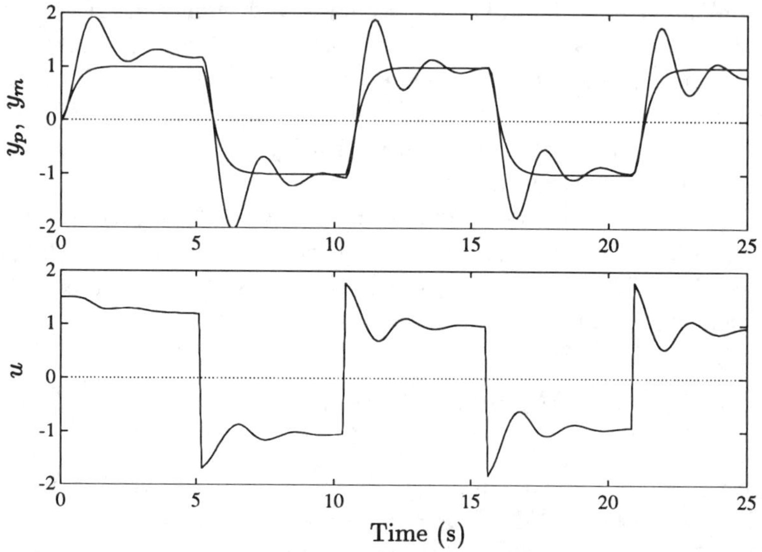

Figure 2.12 Response obtained by using original auxiliary signal generators

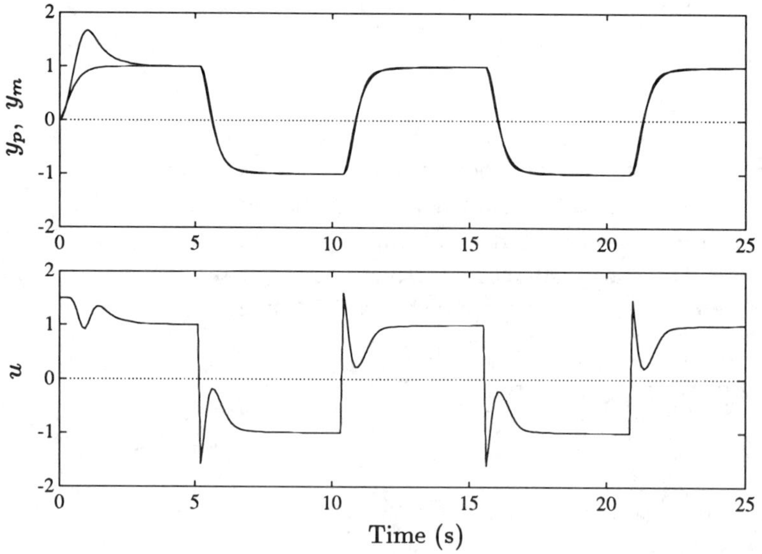

Figure 2.13 Response obtained using a modified second auxiliary signal generator

The filter polynomial L and the ASG numerator polynomial N were chosen as:

$$L = s^2 + 8s + 16 \Longrightarrow W_m L = \frac{128}{s+8}$$
$$N = s^2 + 16s + 64$$

Hence, $W_m L$ is SPR. The adaptation gains for the response using the original ASGs are:

$$\gamma_{k_0} = -2$$
$$\gamma_{c_1} = -1$$
$$\gamma_{c_2} = -0.5$$
$$\gamma_{d_0} = -2$$
$$\gamma_{d_1} = -4$$
$$\gamma_{d_2} = -4$$

Larger gains do not improve the system's behaviour, as was the case in the second-order example. When using a modified ASG2, the adaptation gains are selected a factor 2 larger. The results are shown in figures 2.14 and 2.15. Note that perfect model matching will only occur after some time and thus figures 2.14 and 2.15 show a remaining mismatch at $t = 25$ s.

The processes used in both simulations have poles with a low damping ratio. If the transfer function of the process is in itself more stable, the difference between the original and the modified use of the ASGs becomes less significant. This is due to the greater sensitivity of the system's behaviour to the feedback of the extra estimated process state when the process poles are badly damped.

Note that no formal proof is given of a convergence increase due to the ASG modification. However, choice of 'tight' ASGs in the original method involves a close resemblance between one of the ASG outputs and the process output. The adaptation can then hardly observe a difference between these two, and therefore has difficulty in converging to the correct parameter values corresponding to these two elements of the signal vector. In the modified case, all elements of the signal vector have a clearly different meaning, corresponding to the actual internal process states. Therefore, any ambiguity between signals is avoided, which has positive effects on the convergence.

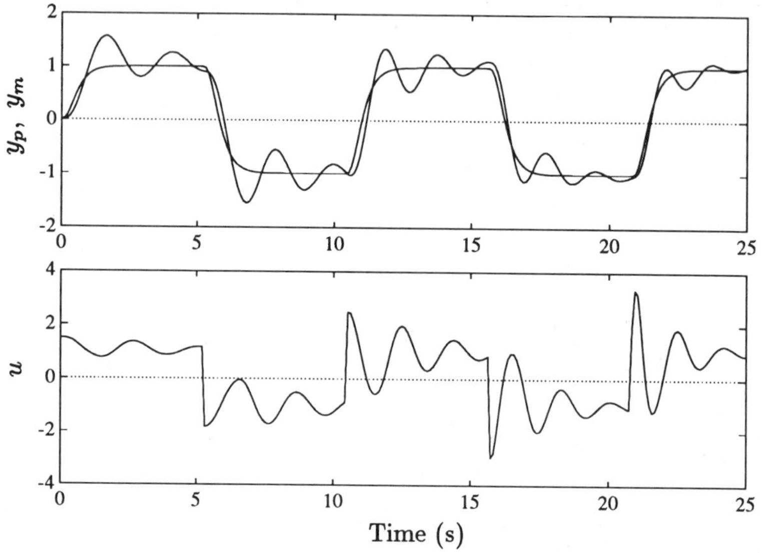

Figure 2.14 Response obtained by using original signal generators and a third-order process

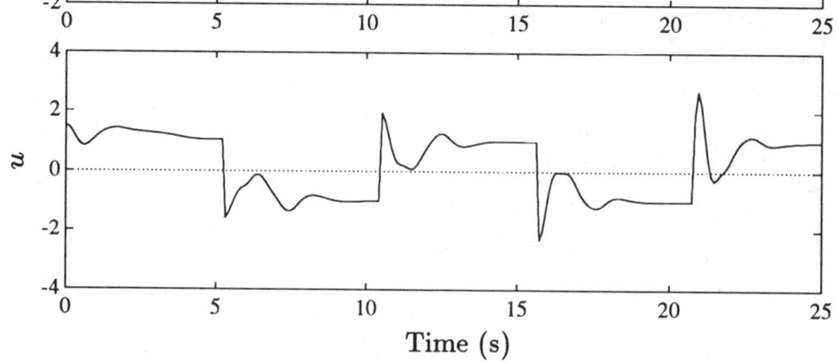

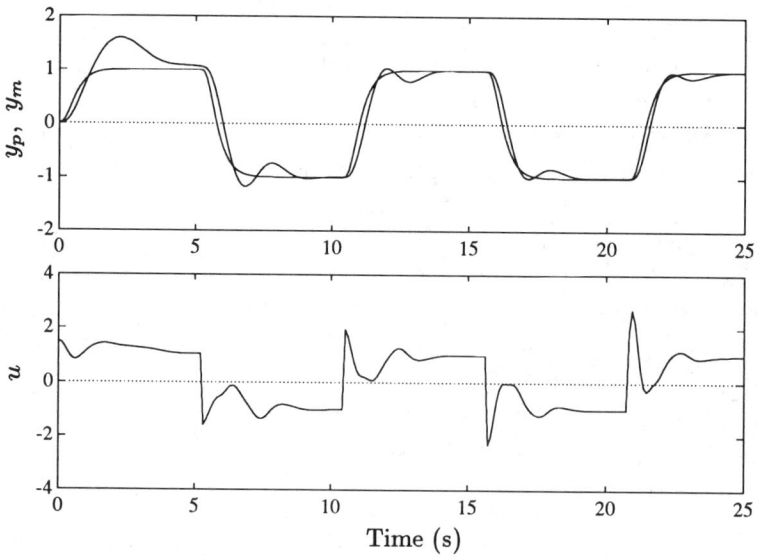

Figure 2.15 Response obtained by using a modified second signal generator and a third-order process

2.7 Tuning aspects

This section describes some tuning aspects related to the augmented error method. The ASGs, the filters L^{-1} and the choice of the adaptation gains will be treated.

The auxiliary signal generators

The auxiliary signal generators (ASGs) play an important role in the augmented error method because they generate the signal vector $\boldsymbol{\omega}$. As mentioned above, the ASGs are of the order $n-1$ and their denominator polynomials must contain the reference model zeros. Usually, this last requirement leaves some freedom in the choice of the ASG poles. For ASG2, the position of these poles determines the accuracy of the state reconstruction, requiring small ASG time constants. However, small time constants allow measurement noise to penetrate into the adaptation mechanism. Many simulations have yielded the following global results.

- The position of the ASG poles affects the convergence considerably, especially if modified ASGs are used.
- Relatively large ASG time constants lead to slow convergence to zero of the output error. The smaller the time constants, the better the convergence because of the greater resemblance between the ASG output vector and the process state. However, choosing the ASG time constants smaller than approximately half those of the process does not improve system behaviour.
- The choice of the ASG time constants is a compromise between accuracy of the state estimation and the presence of measurement noise. They can best be chosen as small as the latter allows, keeping the initial remark in mind.

The filters L^{-1}

The filters L^{-1}, which generate $\boldsymbol{\xi}$ from $\boldsymbol{\omega}$ and in addition are part of the error-augmenting network, play an important role in the stability proof of the augmented error method. The SPR property of the linear part of the error equation depends on them. The filters L^{-1} can take several forms:

- L can be a polynomial in s, and so L^{-1} has only poles and no zeros. The order in this case can be $n-m-1$ or $n-m$.

- L can be a rational transfer function in s. Now, there are both poles and zeros in L^{-1}. The *relative degree* of L^{-1} must be $n-m-1$ or $n-m$.

The last form gives the possibility of choosing $L^{-1} = W_m$, which has implementation advantages. The following remarks result from simulation studies.

- The choice of the filters L^{-1} is not critical as long as the SPR property on $W_m L$ is satisfied (see, however, the last point below).
- The choice $L^{-1} = W_m$ does not generally give very good results. A better choice is to choose L a polynomial of order $n-m-1$. This implies a lower relative order of L^{-1} and thus a smaller phase lag in the filtered signal vector $\boldsymbol{\xi}$, leading to a more exciting signal and so to a better convergence.
- A filter with very small time constants does not necessarily give better results, although the state reconstruction profits from such a choice. The same can be said for filters with large time constants. Time constants in the same range as the reference model time constants seem to be favourable.
- Although $W_m L$ is required to be SPR, a filter which does not meet this requirement (such as a filter with time constants which are too small or which is of a too low relative degree) does not necessarily induce unstable behaviour, and may sometimes even improve the results. The behaviour observed without filter (which usually violates the SPR requirement) may sometimes be even better than those obtained with a theoretically correct filter. This can be explained on the one hand by the observation that the stability theory used is conservative, and so violation of the SPR requirement does not necessarily lead to an unstable system, and on the other hand by the observation that a lower relative degree of L^{-1} implies a smaller phase lag in the signal vector, which is also generally favourable in nonadaptive control.

The adaptation gains

The adaptation gains determine the speed at which adaptation proceeds. This does not, however, imply that higher gains give a better convergence. The following general remarks can be formulated.

- The best choice of the adaptation gains depends very much on the problem at hand. Different problems may require adaptation gains which differ by several orders of magnitude. Simulations are mostly needed to find satisfactory values.

- Not only is the choice of the adaptation gains *themselves* crucial, but *the ratio between them* plays a particularly important role.
- A high value of the adaptation gains may give a larger sensitivity to disturbances, such as unmodelled dynamics. In particular, the relieved SPR requirement obtained by averaging analysis, described in section 1.3.5, is completely based on the assumption of slow adaptation.

2.8 Summary

If all states of the process to be controlled are available, a complete state feedback can serve as primary controller and the SPR property of the linear part of the error equation can be obtained by using a compensated error $\epsilon = \boldsymbol{p}^T \boldsymbol{e}$ in the adaptive law. Extension to multi-input, multi-output processes is straightforward, both in the primary controller design and in the adaptive law derivation. In designing a model reference adaptive controller using only output feedback, two problems arise. First, a more complicated primary controller must be used to satisfy the perfect model-matching condition. Second, the SPR demand on the linear part of the error equation cannot be satisfied by a compensator acting on the state error, because only the output error is available. Several methods have evolved to guarantee an SPR error equation despite this absence.

Landau's method makes use of a full state reconstruction by state-variable filters. By making sure that all corresponding signals have the same phase shift, a stability proof can be achieved. However, Landau's approach is not complete because a primary controller design is not included.

The augmented error method, as described in the literature, provides the designer with a 'standard' scheme for model reference adaptive control using output feedback. The primary controller makes use of 'auxiliary signal generators' which generate extra signal vectors from the process input and output. The SPR property is achieved through an error-augmenting network, which generates an error augmentation such that the resulting error equation becomes SPR. This network consists of only linear filters.

In the original structure of the signal vector in the augmented error method, both y_p and a filtered y_p are fed back to the process input. This results in slow convergence of the method when the process poles are poorly damped. Modifying the ASGs such that they resemble conventionals state-variable filters, producing a filtered estimation of the process state, improves the behaviour considerably in such cases.

Problems

2.1 In section 2.7 on tuning aspects, it was stated that large adaptive gains do not necessarily yield better results than small gains. In section 2.6 on simulation experiments similar statements were made, and it was mentioned that in a specific case 'increasing the adaptation gains does not improve the convergence of the system'. Can you explain these statements if you consider the action of a pure integrator in a linear feedback loop?

2.2 In order to make the linear part of the error equation in the augmented error method, $W_m L$, SPR, a polynomial L is normally used that must be of order $n-m-1$, $n-m$, or $n-m+1$. Can you explain why the last-mentioned of these three is never applied?

2.3 For the first simulation example in section 2.6, calculate the exact controller parameter vector $\boldsymbol{\theta}^*$

2.4 Consider the feedback system:

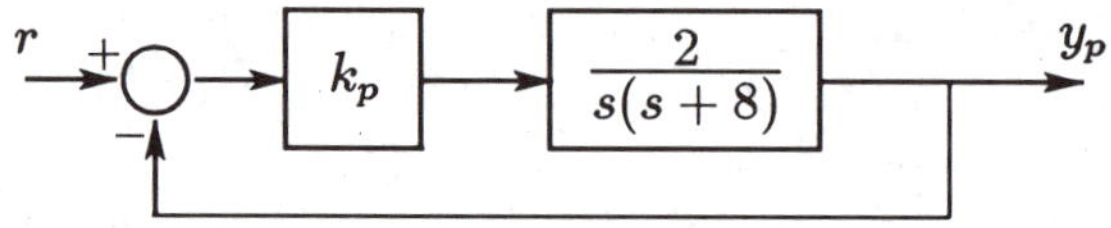

This system is controlled by an MRAC controller which adjusts k_p:

$$\dot{k}_p = -\gamma\epsilon(r - y_p)$$

For this system, derive the error equation for the output error $e = y_p - y_m$. Is the linear part SPR? Design an error-augmenting network to calculate ϵ from e, such that the error equation for ϵ is SPR.

3

Robustness Topics in Adaptive Control

In applying model reference adaptive control to a practical problem, several assumptions are always violated, and the theoretically obtained stability result will generally not hold. An example is the relative order of the reference model numerator and denominator polynomials, which should be equal to the relative degree of the process. In practice, the process order is either not known or may be higher than can be accounted for in the controller. Similarly, the process is not normally disturbance free, and measurement noise or system noise is usually present. It is therefore important to know how the adaptive controller reacts to these imperfections, and how a possibly unsatisfactory behaviour can be improved.

3.1 Introduction: what is robustness?

In recent years, robustness properties in control systems have become a much studied topic (for example, see Kwakernaak, 1988; Ackermann, 1986; Kiendle, 1986). This section introduces the concept of robustness from a nonadaptive point of view, and indicates the difference in 'robustness' in an adaptive context.

Different definitions of robustness are given in the literature. Generally, 'robust control' treats the design of a fixed, linear controller, which is called robust as long as the behaviour of the controlled system remains acceptable if the process deviates from the one the controller was designed for. Thus the controller, being designed for a nominal process W_p, can also properly control the actual process $\overline{W}_p$, which deviates from W_p because of a different parameter set, unmodelled dynamics or external disturbances. Before designing a robust controller, it is necessary to specify the perturbations against which the system must be robust. Usually, an upper bound on the structural uncertainty must be known. Further, the system property that is required to be insensitive to these disturbances must be established (e.g. stability, a zero steady-state error, a small overshoot). Two main classes of the latter property can be distinguished: *stability robustness* and *performance robustness*. In the first class, the emphasis lies on maintaining stability, whereas in the second class the performance is considered. Quantitative measures of robustness, such as the gain margin, are analyzed as a function of the process parameter uncertainty, and hence robustness analysis is clearly linked with sensitivity analysis.

One way to solve the control problem is to make use of frequency-domain methods, for example using a phase-margin and gain-margin analysis. These concepts are generalized by the introduction of singular values. In these methods, a frequency-domain criterion function is minimized off-line. The mathematical space in which this criterion is expressed generally gives the name to the method at hand. The H_∞ and H_2 optimization methods have evolved in this way.

Another way of approaching the problem is to look more directly at the position of the poles of the closed-loop system for various parameter sets. By demanding that these poles stay in a specified area, the controller can be designed. As in MRAC systems, Lyapunov functions can aid in the robustness analysis and can be used to investigate the parameter space for parts that do and do not meet the design specifications (Kiendle, 1986).

The relationship between robust control and adaptive control can be looked upon in different ways. If the structure of the process is known but the parameters are not, an adaptive system can maintain the specifications by adjusting the controller parameters. A fixed controller designed to meet the same specifications, despite the unknown parameters, would normally be more complex (Ortega and

Tang, 1989). Hence, adaptive control can be considered a robustness-improving mechanism, and thus an alternative to robust control. On the other hand, the stability of an adaptive system can be violated by disturbances that would not affect a fixed controller so fundamentally. For example, measurement noise may induce unstable behaviour in a model reference adaptive control system, whereas a fixed controller may suffer from a performance decrease, but not from instability.

Due to these differences in sensitivity of an adaptive controller compared with a fixed controller, the term 'robustness' in an adaptive context has a different interpretation than that in a nonadaptive context. In adaptive control, usually only the *stability* of the controller is considered. Currently, only very limited mathematical tools are available to analyze the *performance* of adaptive systems in the presence of disturbances. In addition, in adaptive control the effect of uncertainty in the process parameters is not of importance for the robustness of the system, because the adaptive controller is designed to be able to cope with this uncertainty. On the other hand, external disturbances and unmodelled dynamics play an important role in the robustness of adaptive control systems. Robustness of adaptive controllers has been a much studied topic (see, for example, Narendra and Annaswamy, 1986a,b,c; Kreisselmeier and Anderson, 1986; Ioannou and Tsakalis, 1986; Kosut, 1986). A good survey of robustness in adaptive control can be found in (Ortega and Tang, 1989).

3.2 Robustness in model reference adaptive control

A robust model reference adaptive system is characterized by the boundedness of all signals in the adaptive loop. Usually, a distinction is made between system behaviour in the presence of external, bounded disturbances and that in the presence of state-dependent disturbances. The first category includes system noise and measurement noise. The second category includes systems in which the process has unmodelled dynamics, which is equivalent to using an adaptive controller of a lower order than is needed to achieve perfect model matching. In this chapter, only bounded disturbances and a limited number of structural disturbances are considered. Unmodelled dynamics are not considered in this chapter, but have a chapter of their own: chapter 5.

This section first presents some tools that can be used to analyze the robustness of model reference adaptive control systems. Next, external disturbances are described and their influence on the error equation is analyzed. Finally, a limited number of other disturbances are considered.

3.2.1 Analysis tools

This section describes some tools which can be used in the robustness analysis of adaptive control systems. First, Lyapunov's stability criterion is reconsidered. Second, the 'persistently exciting' property of signals, which has already been mentioned in chapter 1, is treated and, finally, a stability result for perturbed systems is mentioned.

Stability according to Lyapunov

In model reference adaptive systems the derivative of the Lyapunov function, $\dot{V}(\boldsymbol{x})$, is only semi-negative definite, removing the asymptotic property from the stability. More precisely, if $\boldsymbol{x}^T = (\boldsymbol{e},\ \boldsymbol{\phi})^T$, $\boldsymbol{e} = 0$ is asymptotically stable, whereas $\boldsymbol{\phi} = 0$ is only stable. A sufficient condition for $\boldsymbol{\phi} = 0$ to be *asymptotically* stable is the replacement of $\dot{V}(\boldsymbol{x}) < 0$ by two new conditions:

$$\dot{V}(\boldsymbol{x}) \leq 0 \qquad \text{(Negative } \textit{semi}\text{-definite)}$$

$$\int_t^{t+T} \dot{V}(\boldsymbol{x}(\tau),\tau)\,\mathrm{d}\tau \leq -\gamma(||\boldsymbol{x}(t)||) < 0 \qquad \forall\, t \geq 0 \text{ and } T > 0 \tag{3.1}$$

In these equations, γ is a positive, increasing function with $\gamma(0) = 0$. The second condition makes a semi-negative definite $\dot{V}$ acceptable, but demands that the decrease of the Lyapunov function V in a time interval T must be larger if $||\boldsymbol{x}||$ is larger. Hence, the conditions in equation (3.1) are more strict than the original requirement, because over an interval T, V is required to decrease by an amount specified by $\gamma(\boldsymbol{x})$, rather than any value larger than zero. The progress of a Lyapunov function such as that shown in figure 3.1 may now be acceptable (note that at $t = t_1$, $\dot{V} = 0$). Unfortunately, practical use of the above-mentioned conditions is difficult because analysis of $\dot{V}$ along a trajectory $\boldsymbol{x}(t)$ is necessary in order to be able to test the conditions. However, if the signal vector $\boldsymbol{\omega}$ in the adaptive system is 'persistently exciting', equation (3.1) is automatically satisfied (Kosut, 1986), which is an important result. The next paragraph is devoted to persistently exciting properties.

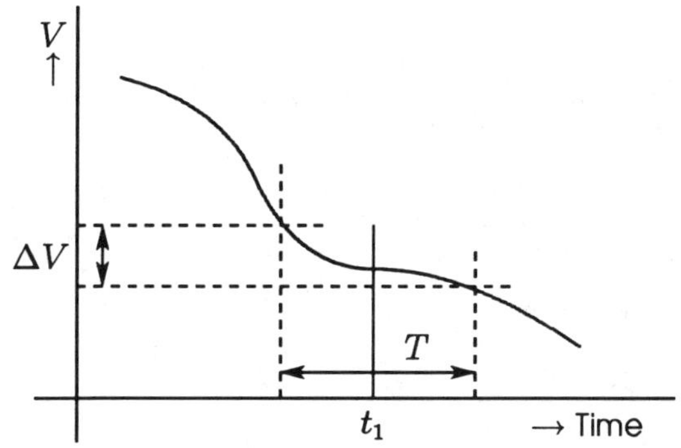

Figure 3.1 Sample progress of Lyapunov function

Persistently exciting signals

The persistently exciting (PE) property of a given signal plays an important role in adaptive systems. As will be illustrated shortly, a persistently exciting signal vector guarantees uniform asymptotic stability (UAS) of both the process-model error $\boldsymbol{e}$ and the parameter error $\boldsymbol{\phi}$, if no system disturbances are present. Several definitions of the PE property have been presented in the literature, some of which are quite complex (see, for example, Kosut, 1986). Generally, the definition of PE is easier to state for discrete-time signals. An important definition is given in (Lee and Narendra, 1988) and can be stated as follows.

Consider a signal vector $\boldsymbol{\omega}(k)$ of dimension n. This signal is *uniformly persistently exciting with degree* μ_0 ($\mu_0 > 0$), if for all $q \geq 0$ there is a positive integer l, such that for all vectors $\boldsymbol{v}$ of dimension n with Euclidian norm $||\boldsymbol{v}|| = 1$, the following holds:

$$\frac{1}{l}\sum_{j=q}^{q+l-1} |\boldsymbol{v}^T\boldsymbol{\omega}(j)| \geq \mu_0 \tag{3.2}$$

This means effectively that for a positive value of μ_0 there is a time interval starting at any instant q, in which no combination of elements of $\boldsymbol{\omega}(k)$ is zero all the time. If equation (3.2) does not hold for every q, the PE property is not uniform.

There is a similar definition for continuous-time signals (Ioannou and Tao,

1989; Sastry and Bodson, 1989). A signal vector $\omega(t)$ is said to be persistently exciting with level μ_0 if there is a $\delta > 0$ such that:

$$\frac{1}{\delta}\int_{t_0}^{t_0+\delta} \omega(\tau)\omega^T(\tau)\,\mathrm{d}\tau \geq \mu_0 \boldsymbol{I}, \quad \forall\, t_0 \geq 0. \tag{3.3}$$

In this equation, $\boldsymbol{I}$ is the identity matrix. This definition is particularly appealing because it is strongly related to the convergence rate in an adaptive system. By rewriting equation (3.3) (Sastry and Bodson, 1989):

$$\frac{1}{\delta}\int_{t_0}^{t_0+\delta} \left[\omega^T(\tau)\boldsymbol{v}\right]^2 \mathrm{d}\tau \geq \mu_0, \ \forall\, t_0 \geq 0 \ \text{ and } \ ||\boldsymbol{v}|| = 1$$

a similar form as in equation (3.2) appears, giving conditions on the energy of ω.

Another definition of PE is given in (Bitmead, 1984), and is used for scalar signals. Here, the signal $y(k)$ to be investigated for PE properties is transformed into a vector $\boldsymbol{x}(k)$ with past elements of $y(k)$:

$$\boldsymbol{x}(k) = [y(k),\ y(k-1),\ \ldots,\ y(k-n+1)]^T$$

Of the vector $\boldsymbol{x}(k)$, a correlation matrix $\boldsymbol{R}$ of order $n \times n$ is calculated:

$$\boldsymbol{R} = E\left\{\boldsymbol{x}(k)\boldsymbol{x}^T(k)\right\} \tag{3.4}$$

The rank m of $\boldsymbol{R}$ is denoted the degree of persistent excitation, and if $y(k)$ is periodic, m denotes the number of spectral lines in $y(k)$. For example, if $y(k)$ consists of a DC component with one sine function, $m = 3$ while a sine function contributes two spectral lines. This measure is different from the parameter μ_0 in the first definition, but a relationship exists between the two. A signal vector ω of dimension n which satisfies the first definition for some $\mu_0 > 0$ is generated by a reference signal r which has a degree $m \geq n$ in the second definition.

Now, for an undisturbed system it can be shown (Lee and Narendra, 1988) that both equilibria $\boldsymbol{e} = 0$ and $\boldsymbol{\phi} = 0$ in a model reference adaptive system are asymptotically stable if the degree of PE of the signal vector $\mu_0 > 0$, or alternatively if the degree of PE m of the reference signal at least equals the dimension of the signal vector. The latter condition shows that a larger number of parameters to be adjusted needs a larger degree of PE of the reference signal. In particular, for identifying n adjustable parameters, a degree of PE $m \geq n$ is needed. For example, if the reference signal r is constant for $-\infty < t < \infty$,

$m = 1$ and so only one parameter can be estimated correctly. In such a case, if the signal vector ω contains more than one element, the degree of PE μ_0 is zero. For example, if ω consists of the process state x_p, a constant reference signal causes some elements of ω to be zero, and hence the corresponding parameters are not updated in the adaptive law. Therefore, according to equation (3.3), $\mu_0 = 0$ and so no convergence to $\phi = 0$ is achieved.

Example 3.1 Persistent excitation for two parameters

Consider the first-order system of figure 3.2, which consists of a process $1/s$ controlled by two parameters k_0 and d_0. Assume that the reference r is a step signal:

$$r = 0 \qquad t < 0$$

$$r = 1 \qquad t \geq 0$$

Such a step signal has a degree of PE of 1, while a constant signal has only one

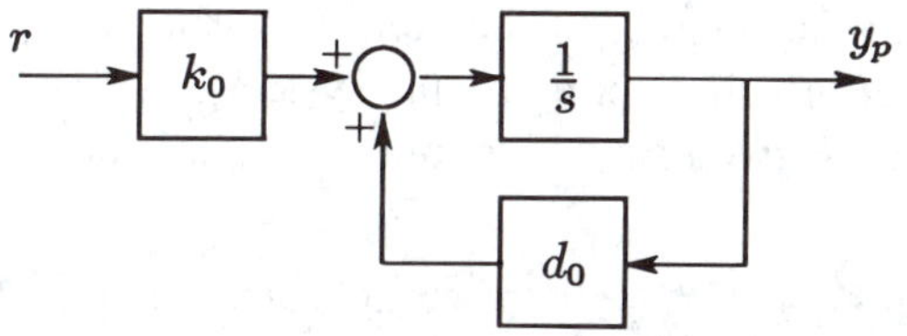

Figure 3.2 Example 3.1, *first-order process*

spectral line, namely for $\omega = 0$. Hence, only one parameter can be successfully identified. Intuitively this is obvious: if r is constant, the only condition for perfect model matching, $y_p = y_m$, is that the DC gain of the closed-loop system equals that of the reference model. Hence, k_0/d_0 is equal to the reference model DC gain, which leaves an infinite number of combinations of k_0 and d_0 for which the output error is zero.

□

The stated result is achieved through the use of the modified Lyapunov stability result, discussed above. The first definition of PE for continuous-time signals (3.3) can be used in the stability analysis of perturbed systems, as will be explained in section 3.2.2. The requirements on the frequency content of r, as expressed in equation (3.4), have led to the terms 'sufficiently rich' (Kreisselmeier and

Narendra, 1982) and 'dominantly rich' (Ioannou and Tao, 1989). In identification, similar results are available (Åström and Wittenmark, 1989).

Summarizing:

- Because $\dot{V}$ is only semi-negative definite the UAS of $\boldsymbol{\phi} = 0$ cannot be guaranteed.
- The UAS of $\boldsymbol{\phi} = 0$ can be guaranteed using integral properties of $\dot{V}$, of which practical application is problematic because the additional inequality is difficult to test.
- If the signal vector $\boldsymbol{\omega}$ is PE of degree $\mu_0 > 0$, or alternatively the degree m of the reference signal is at least the number of adjustable parameters, the solution $\boldsymbol{\phi} = 0$ is UAS. The integral property on $\dot{V}$ is then automatically satisfied.

Stability properties of perturbed systems

If the stability properties of a system $R : \dot{\boldsymbol{x}} = \boldsymbol{f}(\boldsymbol{x}, t)$ are known, some things can be said about the stability of the perturbed system $R_p : \dot{\boldsymbol{x}} = \boldsymbol{f}(\boldsymbol{x}, t) + \boldsymbol{g}(\boldsymbol{x}, t)$ ($\boldsymbol{g}(\boldsymbol{x}, t)$ is the perturbation acting on the system). A link between the stability of R and that of R_p is given by the definition of total stability:

> Assuming that $\boldsymbol{g}(0, t) = 0$, the equilibrium $\boldsymbol{x} = 0$ of R is called totally stable if for every $\epsilon > 0$ two positive numbers, $\delta_1(\epsilon)$ and $\delta_2(\epsilon)$, exist such that if the initial state at $t = t_0$: $||\boldsymbol{x}_0|| < \delta_1$; and the disturbance: $||\boldsymbol{g}(\boldsymbol{x}, t)|| < \delta_2$, the state $\boldsymbol{x}(t, \boldsymbol{x}_0, t_0)$ of R_p evolves such that for every time t, $||\boldsymbol{x}(t, \boldsymbol{x}_0, t_0)|| < \epsilon$.

Thus, the perturbed system can always return to an arbitrarily small vicinity ϵ of the equilibrium $\boldsymbol{x} = 0$ if requirements regarding the upper bound on the initial condition $\boldsymbol{x}_0$, and the magnitude of the perturbation $\boldsymbol{g}(\boldsymbol{x}, t)$, are satisfied. If the unperturbed system R is uniformly asymptotically stable, it is also totally stable.

This result is in accordance with the results regarding PE properties as described above. For the adaptive system to be asymptotically stable, the degree μ_0 of PE must be larger than 0. As will be shown in section 3.2.2, a perturbed system is still stable if the perturbation is smaller than some value which is linearly dependent on μ_0. In other words, every $\mu_0 > 0$ allows some perturbation while still maintaining stability, which accords with the above definition.

3.2.2 External disturbances

This section studies robustness in the presence of external disturbances which are bounded in nature. In a state-space representation, the process equations can be written as:

$$\begin{aligned} \dot{\boldsymbol{x}}_p &= \boldsymbol{A}_p\boldsymbol{x}_p + \boldsymbol{b}_p u + \boldsymbol{D}_p\boldsymbol{\nu}_1 \\ y_p &= \boldsymbol{c}^T\boldsymbol{x}_p + \nu_2 \end{aligned}$$

In these equations, $\boldsymbol{\nu}_1$ is system noise and ν_2 is measurement noise. Both are assumed to be bounded. Both disturbances together can be described in a combined way, denoting ν as the disturbance on the process output in open loop (see also equation (2.18)):

$$y_p = W_p\left[\boldsymbol{\theta}^T(\boldsymbol{\omega}+\boldsymbol{\omega}^\nu)\right] + \nu(t) = W_p\left(\boldsymbol{\theta}^T\boldsymbol{\omega}\right) + \nu'(t) \tag{3.5}$$

with:

$$\nu' = W_p\left(\boldsymbol{\theta}^T\boldsymbol{\omega}^\nu\right) + \nu$$

In equation (3.5), $\boldsymbol{\omega}^\nu$ denotes the part in the signal vector that is completely due to the penetration of the disturbance ν, and hence the actual signal vector consists of the original $\boldsymbol{\omega}$, and $\boldsymbol{\omega}^\nu$. The combined disturbance $\nu'(t)$ depends on $\boldsymbol{\nu}_1(t)$, $\nu_2(t)$, and the process and controller parameters. If the closed-loop system is stable, $\nu'(t)$ is bounded because $\boldsymbol{\nu}_1(t)$ and $\nu_2(t)$ are bounded. The new output error e_1 becomes:

$$\begin{aligned} e_1 &= y_p - y_m \\ &= W_p\left[\boldsymbol{\theta}^T(\boldsymbol{\omega}+\boldsymbol{\omega}^\nu)\right] + \nu - W_m r \\ &= W_p\left(\boldsymbol{\theta}^T\boldsymbol{\omega}\right) - W_m r + W_p\left(\boldsymbol{\theta}^T\boldsymbol{\omega}^\nu\right) + \nu \\ &= W_m\left(\boldsymbol{\phi}^T\boldsymbol{\omega}\right) + \nu' \end{aligned}$$

Now, if no error augmentation is used, in the adaptive law e_1 is multiplied by the signal vector $\boldsymbol{\omega}$:

$$\dot{\boldsymbol{\theta}} = -\boldsymbol{\Gamma} e_1 \boldsymbol{\omega}$$

If the signal vector and the output error are disturbed by ν, this becomes:

$$\begin{aligned}\dot{\boldsymbol{\theta}} &= -\boldsymbol{\Gamma}\left[W_m\left(\boldsymbol{\phi}^T\boldsymbol{\omega}\right)+W_p\left(\boldsymbol{\theta}^T\boldsymbol{\omega}^\nu\right)+\nu\right]\cdot\left[\boldsymbol{\omega}+\boldsymbol{\omega}^\nu\right] \\ &= -\boldsymbol{\Gamma}\left\{\boldsymbol{\omega}W_m\left(\boldsymbol{\phi}^T\boldsymbol{\omega}\right)+\boldsymbol{\omega}\left[W_p\left(\boldsymbol{\theta}^T\boldsymbol{\omega}^\nu\right)+\nu\right]\right. \\ &\quad \left.+\,\boldsymbol{\omega}^\nu W_m\left(\boldsymbol{\phi}^T\boldsymbol{\omega}\right)+\boldsymbol{\omega}^\nu\left[W_p\left(\boldsymbol{\theta}^T\boldsymbol{\omega}^\nu\right)+\nu\right]\right\} \end{aligned} \tag{3.6}$$

Let us assume that ν is stochastic with zero mean. Then, in equation (3.6) the following can be observed. The first term is equal to the original term which occurred in the absence of disturbances. The second and third terms in equation (3.6) consist of a multiplication of stochastic terms produced by ν and the original deterministic terms. The expectation of these terms is zero because the correlation between the stochastic term ν and the deterministic signals in the system is zero. The last term in equation (3.6), however, contains quadratic terms of the disturbance ν. In particular, the signal vector usually contains the disturbed process output $y_p+\nu$, and hence one element of $\boldsymbol{\omega}^\nu$ equals ν. So, in the adaptive law for the element of $\boldsymbol{\theta}$ corresponding to y_p in the signal vector (the feedback factor of y_p in the primary controller), a term $-\nu^2$ arises which is nonzero even if the mean value of ν is zero. Similar problems may occur for the other elements of $\boldsymbol{\theta}$. Even if $\boldsymbol{\theta}=\boldsymbol{\theta}^*$ and the original output error is zero, the parameters will therefore not remain constant due to these nonzero offset terms. This phenomenon is called *drift*.

If error augmentation is used, the result is slightly different because the error-augmenting signal is also affected by ν, and in addition a filtered version of $\boldsymbol{\omega}$, $\boldsymbol{\xi}$, is used. In this case, the nonzero term in the adaptation integrator inputs is:

$$-\boldsymbol{\Gamma}\left(L^{-1}\boldsymbol{\omega}^\nu\right)\left[W_p\left(\boldsymbol{\theta}^T\boldsymbol{\omega}\right)+W_mL\left(\boldsymbol{\phi}L^{-1}-L^{-1}\boldsymbol{\phi}\right)^T\boldsymbol{\omega}^\nu+\nu\right]$$

Considering the element in $\boldsymbol{\theta}$ that corresponds to y_p, part of the cross term now works out as $\nu L^{-1}\nu$. Hence, ν is multiplied by a filtered version of ν, preventing problems if ν is zero-mean white noise. In general however, ν is not white noise, and therefore the problem is similar to that without error augmentation. Note that if the mean value of ν is nonzero, the second and third terms in equation (3.6) do not vanish and hence the problem is more severe.

The parameter drift described above occurs mainly if the original integrator input vector $-\boldsymbol{\Gamma}\,\epsilon\,\boldsymbol{\xi}$ is 'small', or, more formally stated, if the degree of PE of $\boldsymbol{\xi}$ is low. This implies that the largest disturbance effects are expected if the disturbance is relatively large compared to the signal vector $\boldsymbol{\xi}$. The actual disturbance effect depends on the type of disturbance (i.e. the noise colouring).

If the reference signal has too low a degree of PE, there is not enough information to identify all parameters, and an ambiguity remains in the parameter setting. The measurement noise then causes the parameters to drift over a line (or plane) in the parameter space, such that the original error is not affected. Hence, the offset term in the adaptation cannot be cancelled by $\boldsymbol{\Gamma} \epsilon \boldsymbol{\xi}$, and so the drift is unbounded. If the degree of PE of the reference signal is larger, the noise may lead to a wrong parameter setting but could still result in a bounded parameter vector. A formal explanation of these effects can be obtained by using the following result regarding requirements on the degree of PE of the signal vector.

As stated in section 3.2.1, the nonperturbed system is UAS under the condition that the signal vector is PE of any degree $\mu_0 > 0$. In the presence of external disturbances this condition is no longer sufficient. In this case the degree μ_0 of PE of the signal vector must comply with $\mu_0 > \rho\nu_0$, in which $|\nu(t)| < \nu_0 \; \forall \; t$, and ρ is a positive constant. Here not UAS, but only stability (S) is guaranteed. The parameter ρ depends in a nontrivial way on the process transfer function and the parameter vector $\boldsymbol{\theta}$. The important result is, however, that now not every $\mu_0 > 0$ is sufficient, but that μ_0 must exceed some threshold (if the adaptive law is not modified). A translation to demands on the reference signal r is, however, difficult. Some schemes exist (Draijer, 1988) that make use of a monitoring algorithm which decides whether the signal vector is exciting enough to continue adaptation, and switches off adaptation otherwise. Care must be taken with these algorithms because they resemble a dead zone in the adaptation, and, as will be described in section 3.3.1, this may yield undesirable effects.

Example 3.2 First-order process disturbed by measurement noise

In this example, we consider a first-order process, controlled by a controller consisting of a feedforward parameter k_0 and a feedback parameter d_0, as illustrated in figure 3.3. This system corresponds with that used in example 3.1, which did, however, not suffer from disturbances. The disturbance ν is assumed to be zero-mean white noise. The controller parameter vector, the signal vector, and

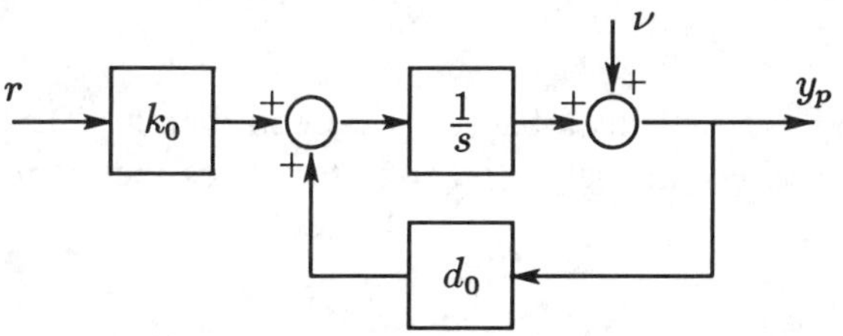

Figure 3.3 Example 3.2, first-order process with measurement noise

the output error are, respectively:

$$\begin{aligned}
\theta^T &= (k_0,\ d_0) \\
\omega^T &= (r,\ y_p) = \left(r,\ y_p' + \nu'\right) \\
e_1 &= y_p - y_m = e_1' + \nu'
\end{aligned}$$

Here, y_p and e_1 denote the actual process output and the output error respectively; ν is the actual disturbance occurring; y_p' is the process output as it would be without disturbance; ν' is the disturbance as it appears at the process output (which differs from ν due to the feedback). Hence, $y_p = y_p' + \nu'$ and e_1 consists of the original error signal e_1' and the disturbance ν'. Applying the 'standard' adaptive laws for k_0 yields:

$$\begin{aligned}
\dot{k}_0 &= -\gamma_0 e_1 r \\
&= -\gamma_0 \left(e_1' + \nu'\right) r \\
&= -\gamma_0 e_1' r - \gamma_0 \nu' r
\end{aligned}$$

Now, the first term is equal to the original term without disturbance. The expectation of the second term is zero, while the correlation between the reference r and the disturbance ν' is zero. Adaptation of k_0 therefore operates correctly. For d_0 the situation is slightly different:

$$\begin{aligned}
\dot{d}_0 &= -\gamma_1 e_1 y_p \\
&= -\gamma_1 \left(e_1' + \nu'\right)\left(y_p' + \nu'\right) \\
&= -\gamma_1 \left[e_1' y_p' + e_1' \nu' + \nu' y_p' + (\nu')^2\right]
\end{aligned}$$

Here, the first term is equal to the original term and the second and third terms are zero. However, the last term $(\nu')^2$ is nonzero although the mean value of ν' is zero, and hence drift in d_0 is likely to occur if $e_1' y_p'$ is relatively small.

□

Previously, only external disturbances were considered. The following section will describe a limited number of other disturbances that play an important role in MRAC systems.

3.2.3 Other disturbances

In addition to external disturbances, structural process perturbations are important, such as:

- Dead time in the process transfer function.
- Process nonlinearities.

Dead time in the process has two important effects on the adaptive scheme. First, the perfect model-matching condition is no longer satisfied. Second, phase differences arise between y_p and y_m, and between the various elements in the signal vector $\boldsymbol{\xi}$ and the error ϵ. These phase differences can be eliminated, if the dead time is known, by placing delay operators in some of the signal flows. However, the violation of the perfect model-matching condition still prohibits a stability proof. Dead time is considered a major problem in MRAC and workable formal solutions are not known. A heuristic solution that can deal with a limited dead time will be presented in section 3.3.3.

Process nonlinearities are another important issue in MRAC. If the nonlinearities are 'invertible', a nonlinear feedback can be designed that linearizes the overall transfer function (Isidori, 1985). The first steps towards adaptive tuning of the parameters in such a nonlinear feedback have been taken (Sastry and Isidori, 1989; Sastry and Bodson, 1989). In this chapter, the nonlinearity is considered a disturbance and so the robustness of the adaptive controller is important in this respect. Usually, nonlinearities generate more negative effects as the number of adjustable parameters becomes larger. Only a very limited number of nonlinearities will be mentioned here.

Limitations on the process input can be dealt with by stopping the adaptation as soon as the limitation occurs. Further, implementing the same limitation in the reference model may be valuable but is normally difficult because the reference model structure does not have a feedback configuration. Another way of solving the problem is to *avoid* the limitation by choosing a proper reference signal (for example, achieved by a series-type reference model, see van Amerongen, 1982). Alternatively, a model-adjustment scheme may prove valuable (chapter 6).

Dead zones in the process transfer normally lead to 'limit cycles' in the adaptation loop, due to the integrating term in the parameter adjustment law. If the dead zone is known, a direct compensation can be implemented.

3.3 Robustness-improving mechanisms

This section describes several ways of improving the robustness of MRAC systems. First, methods of coping with external disturbances are presented, and, next, adaptive offset compensation which deals with process input offsets is described. Finally, a heuristic solution for dead time in the process transfer function is presented.

3.3.1 Modifications of the adaptive law

In this section, several modifications of the adaptive law are discussed that improve the robustness against external disturbances. All modifications have in common that there are theoretical results that guarantee stability if the modification parameters are chosen properly. The modifications can thus be seen as alternatives for requirements on the degree of PE of the signal vector $\boldsymbol{\xi}$, which are difficult to test in practice.

Experiment description

To compare the effectiveness of the modifications, the same simulation experiment was performed for all modifications. As disturbance, coloured measurement noise was used. As described in section 3.2.2, white measurement noise may not impose problems if error augmentation is used, because the disturbance signals that are multiplied in the adaptive law have no correlation. The simulations to test the modifications consist of two parts. The first part of the response is obtained by applying a block input to the system which does not suffer from disturbances. This represents the behaviour of the system in the ideal case. During the second part of the response, the reference input is set to zero and coloured measurement noise is added to the process output. This response can be considered as a test for robustness because parameter drift (instability) is most likely to occur here. Note that, in the second part, the degree of PE of the reference signal is zero. The modification being studied is tuned to this second part of the response. The first part shows the behaviour of the system under disturbance-free conditions. A modification is considered 'good' if the system response is satisfactory both with and without external disturbance.

The process and reference model are equivalent to those used in chapter 2:

$$W_p = \frac{8}{s^2 + 2s + 8} \qquad \text{with poles at } s = -1 \pm 2.65j$$

$$W_m = \frac{16}{s^2 + 8s + 16} \qquad \text{with poles at } s = -4 \;\; (2\times)$$

The measurement noise is obtained by filtering white noise using a filter with a damping ratio $\zeta = 0.1$ and a bandwidth $\omega_n = 25$ rad/s, which results in the frequency spectrum of figure 3.4. Figure 3.5 shows the behaviour of the adaptive

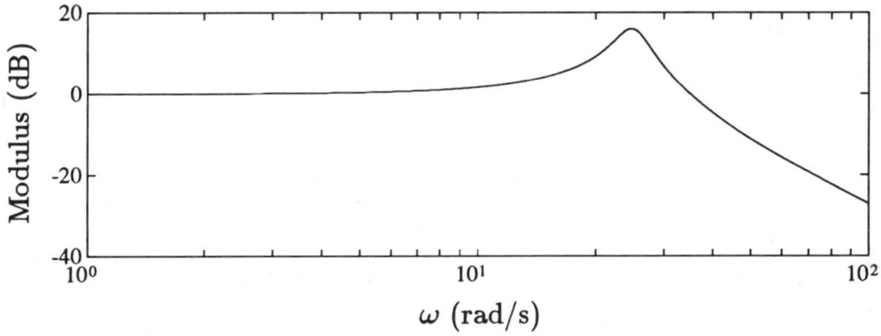

Figure 3.4 Frequency spectrum of measurement noise used in the robustness tests

system when no modification is used. The parameter drift in the second part can clearly be seen. Low-frequency measurement noise appears to impose fewer problems because of the high-pass characteristic of ASG2. To improve the robustness, the following four modifications may be considered.

Dead zone in the adaptation loop

In this method the adaptation is stopped when $|\epsilon| \leq \nu_0$, with ν_0 a known upper bound on the disturbance (Kreisselmeier and Narendra, 1982; Peterson and Narendra, 1982). If $|\epsilon| > \nu_0$ the standard adaptation law is used (here including normalization):

$$\dot{\boldsymbol{\theta}} = -\boldsymbol{\Gamma}\frac{\epsilon\boldsymbol{\xi}}{1 + \boldsymbol{\xi}^T\boldsymbol{\xi}}$$

This method guarantees stability of the complete adaptive system if the magnitude of the dead zone is larger than ν_0. There are several alternative dead zone

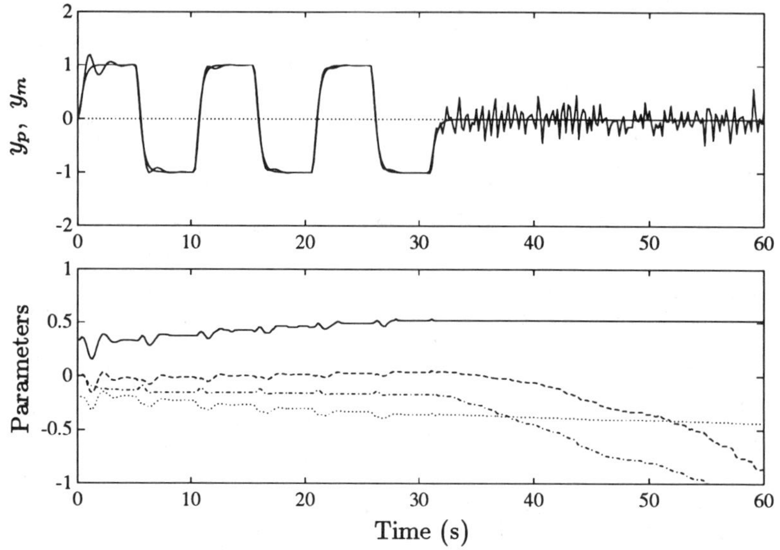

Figure 3.5 Response of second-order process with measurement noise, without modifications of the adaptive law

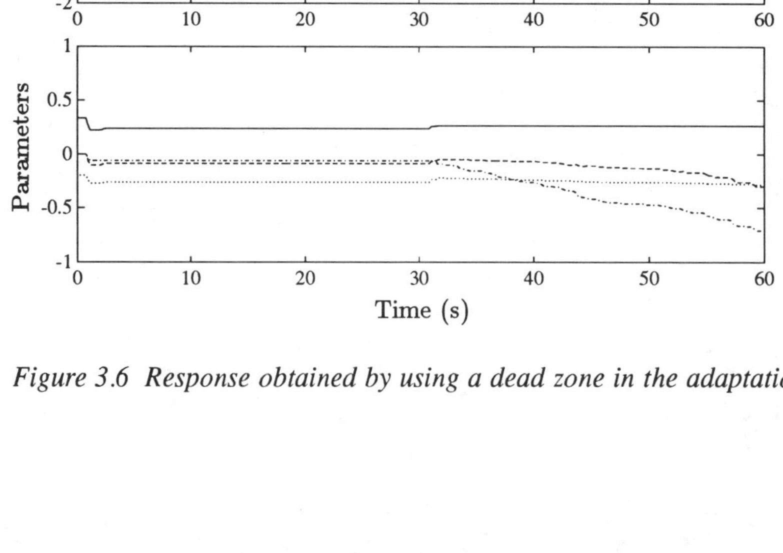

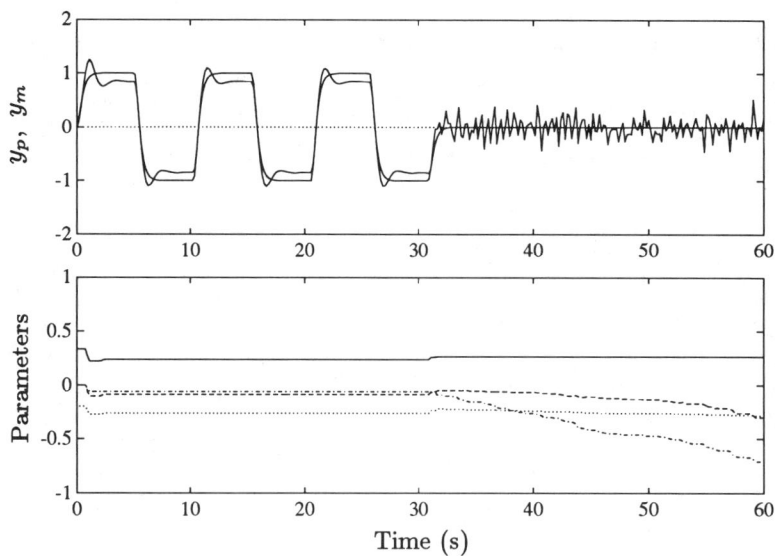

Figure 3.6 Response obtained by using a dead zone in the adaptation

implementations (Kreisselmeier and Anderson, 1986). Introducing a dead zone in the adaptation requires knowledge of an upper bound ν_0 on the disturbance $\nu(t)$, which may not be available. Further, if no disturbances are present, the parameter error $\boldsymbol{\phi}$ will not converge to zero and a steady-state difference between process and model output will remain.

Figure 3.6 shows the effect of using a dead zone in the adaptation. The parameter drift has been reduced but does not vanish completely; apparently the implemented dead zone is not large enough. A steady-state error is introduced when no disturbances are present. Because of this, the dead zone is not considered a good solution for the robustness problem.

Known bound on θ^*

If an upper bound $||\boldsymbol{\theta}^*||_{\max}$ on the norm of the correct parameter setting $\boldsymbol{\theta}^*$ is known, this knowledge can be used in the adaptation (Narendra and Annaswamy, 1986a,b; Kreisselmeier and Narendra, 1982). The method proposed in the latter paper adds an extra term to the equation for $\boldsymbol{\theta}$ as soon as $||\boldsymbol{\theta}|| > ||\boldsymbol{\theta}^*||_{\max}$:

$$\dot{\boldsymbol{\theta}} = -\boldsymbol{\Gamma}\frac{\epsilon\boldsymbol{\xi}}{1+\boldsymbol{\xi}^T\boldsymbol{\xi}} - \boldsymbol{\theta}\left(\frac{||\boldsymbol{\theta}||}{||\boldsymbol{\theta}^*||_{\max}} - 1\right)^2$$

This involves a change from a pure integrating term to a first-order transfer function when $\boldsymbol{\theta}$ moves outside the area in which $\boldsymbol{\theta}^*$ must surely be. The pole of the transfer function moves further to the left as $\boldsymbol{\theta}$ moves further out of this area. Stability of the adaptive system can be proved, but practical application needs knowledge of the bound $||\boldsymbol{\theta}^*||_{\max}$, which may not be available. Besides, no scaling is used for parameters which vary over a wider range than others. An advantage is that in the absence of disturbances the system behaves exactly like the unmodified system. However, parameter drift can occur freely as long as $||\boldsymbol{\theta}|| < ||\boldsymbol{\theta}^*||_{\max}$. A similar modification can be used if a lower bound $||\boldsymbol{\theta}^*||_{\min}$ is known.

Sigma modification

This modification involves an extra term $-\sigma\boldsymbol{\theta}$ in the equation for $\dot{\boldsymbol{\theta}}$:

$$\dot{\boldsymbol{\theta}} = -\boldsymbol{\Gamma}\frac{\epsilon\boldsymbol{\xi}}{1+\boldsymbol{\xi}^T\boldsymbol{\xi}} - \frac{\sigma}{1+\boldsymbol{\xi}^T\boldsymbol{\xi}}\left[\boldsymbol{\theta} - \boldsymbol{\theta}(0)\right], \qquad \sigma > 0$$

The extra term shifts the pole in the adaptive loop from the origin to a fixed location on the negative real axis (Narendra and Annaswamy, 1986a,b). The pole in the left half of the s-plane has the effect that a nonzero error signal is required to retain the parameter values. This means that the augmented error ϵ will not go to zero, even in the absence of disturbances. In this case, the behaviour of the system can be even worse than in the presence of disturbances. Although no explicit knowledge is necessary about the norm of the parameter vector or an upper bound on the disturbance, the behaviour of the system depends heavily on the choice of σ. A large value of σ makes the system insensitive to perturbations, but in the absence of disturbances a small value is needed if a small deviation from the reference model is required. This is illustrated in figure 3.7. The value of σ is chosen such that drift is avoided in the second part of the response. However, if no disturbances are present, 'leakage' of the parameter values occurs. This means that if the element of the state vector, which corresponds to a specific

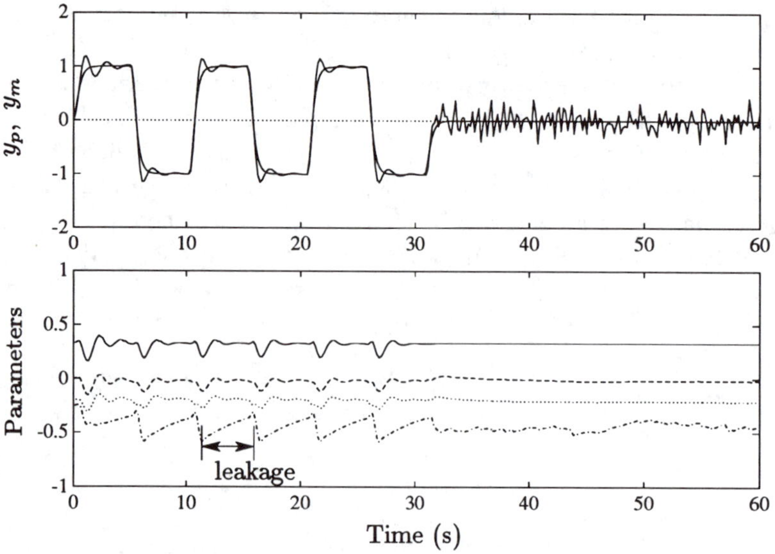

Figure 3.7 Response obtained with σ-modification

parameter, tends to zero, the integrator in the adaptation leaks and the parameter drifts towards its initial value. This can be seen in figure 3.7: during the periods in which the reference signal is constant, parameter d_1 (corresponding to the filtered derivative of y_p in ASG2) tends to zero. Leakage involves a new start-up of the adaptation after a set-point change, which results in a nonvanishing overshoot in figure 3.7.

Gamma modification

As with σ-modification, with γ-modification the integrator is changed to a first-order transfer function (Narendra and Annaswamy, 1986a,b). The position of the pole, however, is continuously adjusted to the magnitude of the error ϵ. The adaptive law becomes:

$$\dot{\boldsymbol{\theta}} = -\boldsymbol{\Gamma}\frac{\epsilon\boldsymbol{\xi}}{1+\boldsymbol{\xi}^T\boldsymbol{\xi}} - \frac{\gamma|\epsilon|}{1+\boldsymbol{\xi}^T\boldsymbol{\xi}}\left[\boldsymbol{\theta} - \boldsymbol{\theta}(0)\right], \quad \gamma > 0$$

In section 3.2.1, it was stated that the system without modification is UAS, and so $\boldsymbol{\phi} \to 0$ if the signal vector $\boldsymbol{\xi}$ is PE with any degree $\mu_0 > 0$. Using γ-modification, the unperturbed system can be guaranteed to be UAS if $0 \leq \gamma \leq \gamma^*$. If γ is larger than γ^*, the same leakage problems as with σ-modification may occur. The upper bound γ^* depends on the convergence properties of the system and thus, among others, on the degree of PE μ_0 of $\boldsymbol{\xi}$. The larger μ_0, the larger the allowed area for γ for which the parameters will converge to their true values. If the parameter γ is fixed, there may be a minimal value of μ_0 (> 0) below which the system will not converge to $\boldsymbol{\phi} = 0$. Because μ_0 depends on the reference signal r, the process response, and the transfer function of the auxiliary signal generators, no explicit conditions on r can be established. However, it is clear that in the absence of disturbances γ-modification places heavier demands on the reference signal than does the unmodified system.

In addition to this upper bound on γ, which in effect protects the adaptation against leakage, there is a lower bound if disturbances are present. Because $\gamma = 0$ does not suffice in this case, it is clear that γ must exceed some threshold which is a function of the disturbance magnitude. Simulations show that too large a γ inhibits convergence (as σ-modification does), but too small a γ is insufficient to avoid drift. The choice is, however, not very critical, as shown in figure 3.8: in both parts of the response the modification suffices, despite the different circumstances.

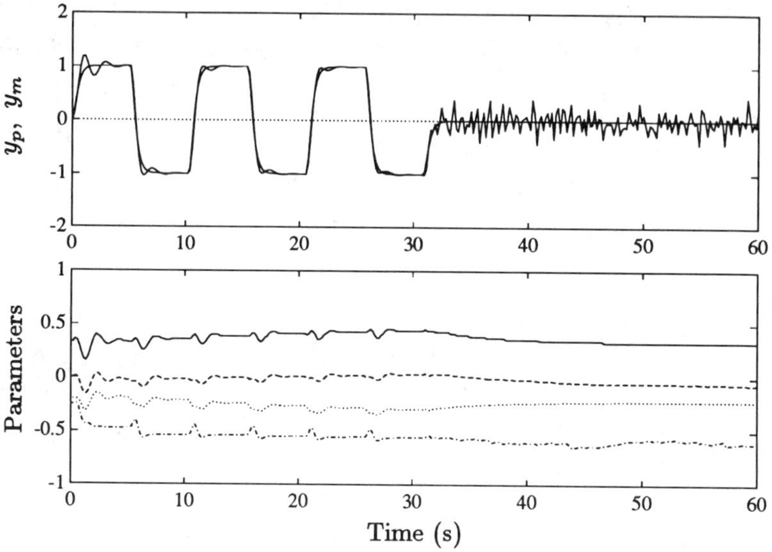

Figure 3.8 Response obtained with γ-modification

3.3.2 Deterministic input disturbances

Deterministic input disturbances, like input offset, formally belong to the class of external disturbances, as described in section 3.2. The modifications of the adaptive law, given in section 3.3.1, can all be used for this type of disturbance. Although stability is guaranteed, these do not yield a satisfactory performance. This is due to the fact that at a certain steady-state process output there is a parameter setting for which $y_p = y_m$. However, for different set points the correct parameter tuning differs, and so no convergence can take place if the reference signal varies.

The problem of input offset can be solved by adding an extra element '1' to the signal vector $\boldsymbol{\omega}$ which serves as an extra input to the process via an extra adjustable parameter k_1. This parameter is updated in the same way as the others:

$$\dot{k}_1 = -\gamma_{2n+1}\frac{\epsilon \cdot 1}{1 + \xi^T \xi}$$

The use of adaptive offset compensation in the augmented error method leads to the primary controller structure of figure 3.9. The positive effects of this approach are shown in figures 3.10 and 3.11. In figure 3.10, an input offset of 0.2

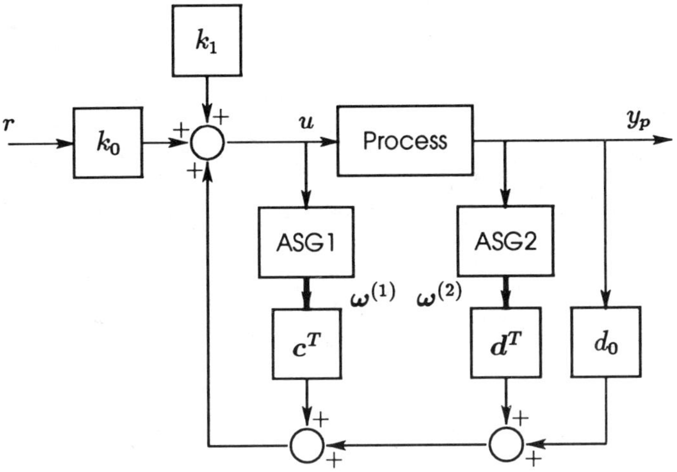

Figure 3.9 Primary controller structure of the augmented error method, expanded with offset compensation

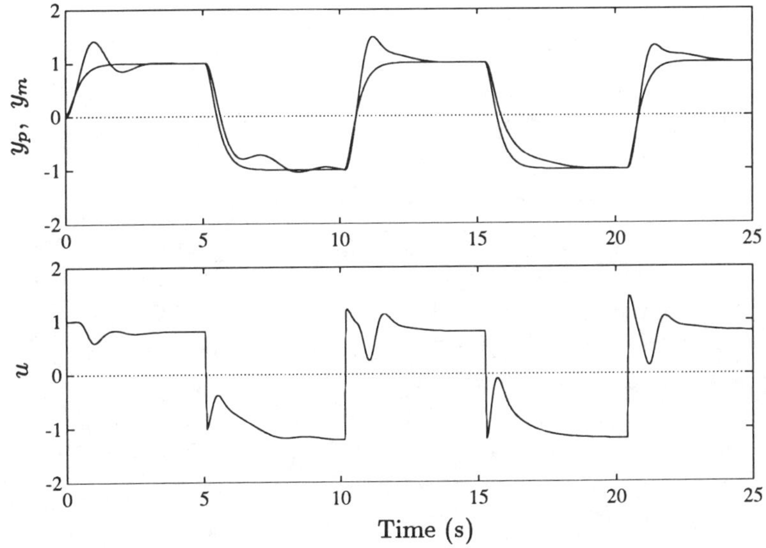

Figure 3.10 Response of second-order process with an input offset of 0.2, *obtained without offset compensation*

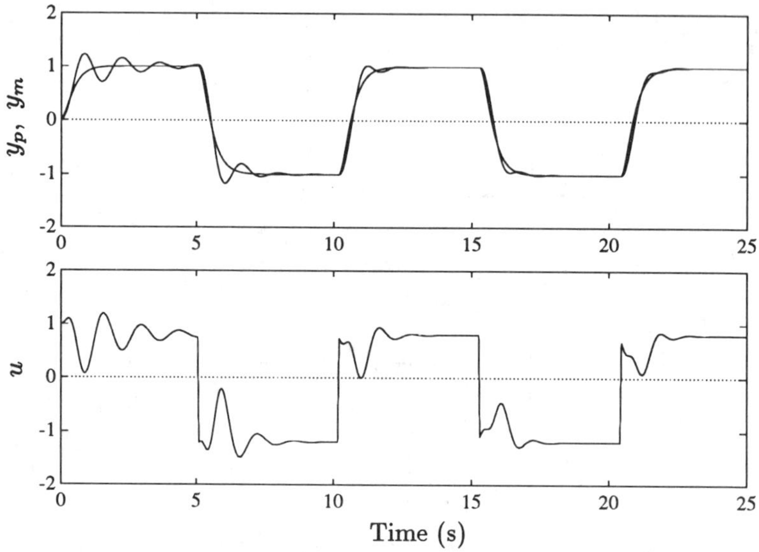

Figure 3.11 Response of the same process, obtained by using adaptive offset compensation

is added to the process input, while no offset compensation is used. Figure 3.11 shows the result with adaptive offset compensation.

Example 3.3 Lyapunov analysis of offset compensation

In this example, we consider the first-order system of example 1.2, which in this case suffers from a constant input disturbance δ:

$$y_p = \frac{b}{s+1}u + \frac{b}{s+1}\delta$$

The model is the same as in example 1.2:

$$y_m = \frac{1}{s+1}r$$

The control u incorporates an offset-compensating term:

$$u = k_0 r + k_1$$

Hence, the process and model differential equations yield a new error equation:

$$
\begin{aligned}
\dot{y}_p &= b\,(k_0 r + k_1 + \delta) - y_p \\
\dot{y}_m &= r - y_m \\
\dot{e} = \dot{y}_p - \dot{y}_m &= (bk_0 - 1)\,r + b\,(k_1 + \delta) - y_p + y_m \\
&= -e + (bk_0 - 1)\,r + b\,(k_1 + \delta)
\end{aligned}
$$

Now, the parameter error is made up of two parts, $(bk_0 - 1)$ and $b(k_1 + \delta)$. The new Lyapunov function incorporates both parameter errors:

$$
V = e^2 + \frac{1}{\gamma_0}\,(bk_0 - 1)^2 + \frac{1}{\gamma_1}\,(bk_1 + b\delta)^2
$$

Differentiating V, following the lines of example 1.2, yields:

$$
\begin{aligned}
\dot{V} &= 2e\dot{e} + \frac{2b}{\gamma_0}\dot{k}_0\,(bk_0 - 1) + \frac{2b}{\gamma_1}\dot{k}_1\,(bk_1 + b\delta) \\
&= -2e^2 + 2e\,(bk_0 - 1)\,r + 2eb\,(k_1 + \delta) + \frac{2b}{\gamma_0}\dot{k}_0\,(bk_0 - 1) + \frac{2b}{\gamma_1}\dot{k}_1 b\,(k_1 + \delta)
\end{aligned}
$$

Putting the last four terms to zero:

$$
\begin{aligned}
2e\,(bk_0 - 1)\,r + \frac{2b}{\gamma_0}\dot{k}_0\,(bk_0 - 1) &= 0 \\
2eb\,(k_1 + \delta) + \frac{2b}{\gamma_1}\dot{k}_1 b\,(k_1 + \delta) &= 0
\end{aligned}
$$

directly yields the adaptive laws:

$$
\begin{aligned}
2er + \frac{2b}{\gamma_0}\dot{k}_0 = 0 \quad &\Longrightarrow \quad \dot{k}_0 = -\frac{\gamma_0}{b}e \cdot r \\
2e + \frac{2b}{\gamma_1}\dot{k}_1 = 0 \quad &\Longrightarrow \quad \dot{k}_1 = -\frac{\gamma_1}{b}e \cdot 1
\end{aligned}
$$

The adaptive law for k_1 has the same form as that for k_0 and guarantees stability of the adaptive system.

□

Example 3.4 Offset compensation in the method using state feedback

In this example we consider the same constant input disturbance δ as in example 3.3 for a process which is controlled by state feedback. If state feedback is applied, as described in section 2.1, the offset compensation requires the following modification of the control signal u:

$$u = k_0 r + \boldsymbol{k}_b^T \boldsymbol{x}_p + k_{n+1}$$

The parameter k_{n+1} can be regarded as controlling a second input with a constant value 1, in addition to the existing reference r. Hence, the new closed-loop control matrix $\boldsymbol{b}_c$, and the reference model control matrix $\boldsymbol{b}_m$, become:

$$\boldsymbol{b}_c = \begin{pmatrix} 0 & 0 \\ 0 & 0 \\ b_{pn} k_0 & b_{pn} k_{n+1} + b_{pn}\delta \end{pmatrix}$$

$$\boldsymbol{b}_m = \begin{pmatrix} 0 & 0 \\ 0 & 0 \\ b_{mn} & 0 \end{pmatrix}$$

Here, the input vector is $(r,\ 1)^T$. Adding extra elements to $\boldsymbol{\theta}$ and $\boldsymbol{\omega}$:

$$\boldsymbol{\theta}^T = \left(k_0,\ \boldsymbol{k}_b^T,\ k_{n+1}\right)$$

$$\boldsymbol{\omega}^T = \left(r,\ \boldsymbol{x}_p^T,\ 1\right)$$

leads to exactly the same error equation as was presented in section 2.1, in turn yielding the same adaptive laws for all parameters. The adaptive law for the new parameter k_{n+1} becomes:

$$b_{pn}\dot{k}_{n+1} = -\gamma_{n+1}\left(\boldsymbol{p}^T \boldsymbol{e}\right) \cdot 1$$

$$\Longrightarrow \dot{k}_{n+1} = -\gamma'_{n+1}\left(\boldsymbol{p}^T \boldsymbol{e}\right) \cdot 1$$

Note that in this example, and in example 3.3, δ was assumed to be a constant input disturbance. In practice, δ may vary slowly with time, the allowed variation rate depending on the adaptation gain for the adaptive offset compensation. In many practical applications, the offset compensation is actually used to compensate minor process nonlinearities (operating-point differences).

□

In the above, only a constant δ is considered. However, any δ of which the form is known but the amplitude is not can be compensated for by this procedure. Suppose, for example, that the input is disturbed by $\delta = A \sin(\omega t)$, in which A is the (unknown) amplitude of the disturbing sinusoid. In this example, an extra signal $k \cdot \sin(\omega t)$ can be added to the control signal u. Adaptation of k is performed in the same way as before:

$$\dot{k}_1 = -\gamma_{2n+1} \frac{\epsilon \cdot \sin(\omega t)}{1 + \xi^T \xi}$$

for the augmented error method, and

$$\dot{k}_{n+1} = -\gamma'_{n+1} \left(p^T e \right) \cdot \sin(\omega t)$$

for the method using state feedback. Note that only *input* disturbances are considered. Disturbances acting on other parts of the process are more complicated. When a method using only input and output measurements is applied (such as the augmented error method), it suffices to determine what kind of compensation on u is needed to cancel the effect of the disturbance on the process output. However, if the method using state feedback is applied, process and model states may not be comparable due to the disturbance occurring, and application of the above procedure may be impossible.

3.3.3 Making use of an orthogonal error signal

The presence of dead time in the process is usually a problem in MRAC. On the one hand, this is caused by phase differences in various signals to be compared and multiplied, and more generally by the phase lag occurring in the adaptive loop. On the other hand, the primary controller is not equipped to deal with dead time, and hence perfect model matching is not possible. Because of these two effects, dead time may easily introduce instability. A generally applicable solution of this problem will not be given in this section, but a heuristic approach will be presented which is based on the following observation.

Suppose the process and model outputs at a certain time t look like those in figure 3.12. It can be seen that, especially during the transient, the dead time in the process output causes a large output error signal (a). This does not correspond with the visual observation that the difference between the process and model outputs is not so large at all. The large error signal occurs because of the steepness of the

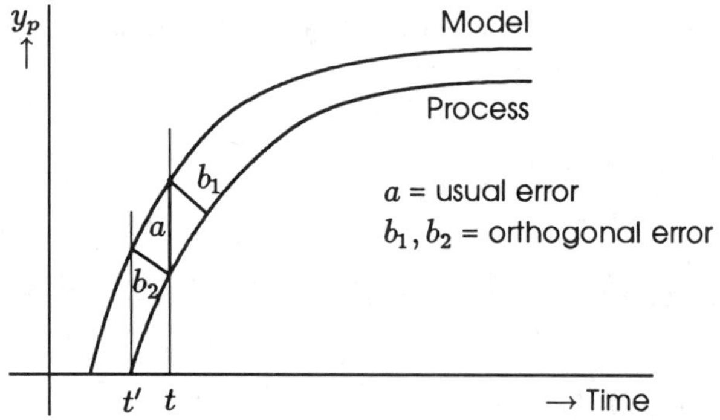

Figure 3.12 Process and model outputs after set-point change if the process suffers from dead time

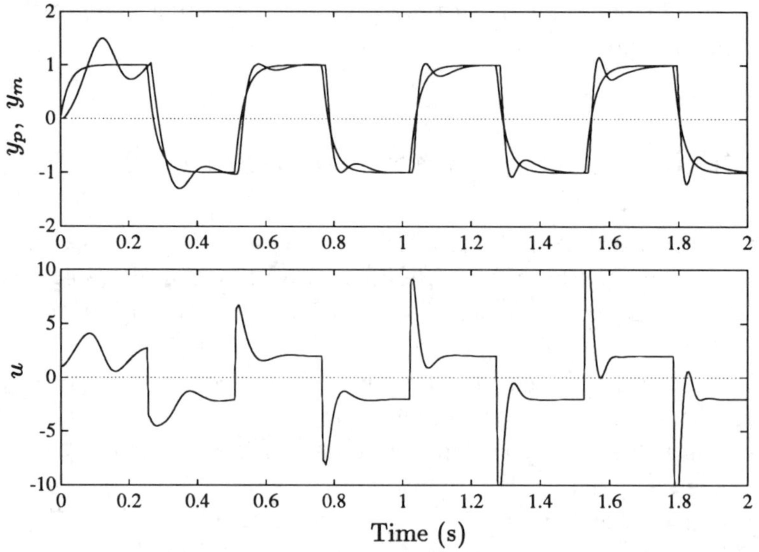

Figure 3.13 Response obtained with first-order process with dead time

responses, which has the effect that a relatively small time delay results in a large difference between the model output and the process output. Figure 3.13 shows the effects that the time delay has on the adaptation mechanism for a first-order process with dead time:

$$W_p = \frac{0.5e^{-0.01s}}{0.05s+1}, \qquad W_m = \frac{1}{0.025s+1}$$

Because of the large error signal during the transients, the parameters are updated in an incorrect way during these transients. Later on, when a steady state is to occur, the incorrect parameter values must be adjusted for this initial misadaptation. In time, this leads to instability.

As mentioned before, although the output error during a transient is large, a visual interpretation shows a much smaller difference between process and model outputs. Intuitively, it seems better to choose an error signal between the two outputs which is perpendicular to the reference model output at time t, as shown in figure 3.12 (b_1). This signal is called the 'orthogonal error' because it has a geometric direction which is orthogonal with respect to the model output. The orthogonal error gives a better heuristic measure for the actual process-model error.

The calculation of an 'orthogonal error signal' instead of the conventional error signal is accomplished as follows. At the current time t, the reference model output and its derivative are available, and thus the perpendicular to the model output is completely determined. Because past values of the process output are known and stored in memory, the intersection of this perpendicular with the process output can be found *if this intersection takes place in the past.* This would be the case if, in figure 3.12, the process and model responses were reversed. In that case, the perpendicular b_2 would be found to intersect the process output at time t'. The length of b_2 is then used as the 'orthogonal error'.

However, the situation is bound to occur where the intersection of the perpendicular with the process output takes place *in the future*, as shown in figure 3.12 (b_1). In this situation, the length of perpendicular b_1 cannot be determined because future process outputs are not available. To deal with this problem, there are two possible solutions. First, it is possible to determine the perpendicular to the *process* output, and determine the intersection with the past reference model output. This method would, however, require knowledge of the derivative of the process output, which may not be available. In addition, measurement noise would disturb the algorithm. Therefore, another approach is applied which still makes use of the derivative of the reference model output, which *is* available. This approach is illustrated in figure 3.14. At time t, a search for a perpendicular on the model output is initiated, going 'backwards' from time t into the past. For past values

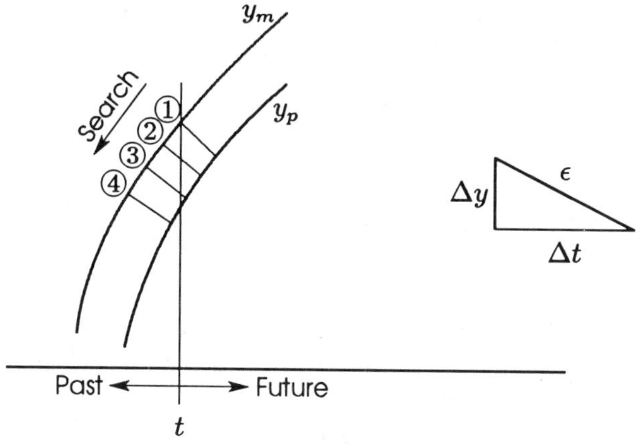

Figure 3.14 Search procedure for orthogonal error

of y_m, the perpendicular on y_m is calculated and tested for an intersection with the process output until time t. In this way, in figure 3.14 the perpendicular lines denoted by ①, ②, ③ and ④ are evaluated. The first three evaluations are not successful: no intersection is found of the perpendicular with a known process output value. The fourth evaluation is successful and is used as the 'orthogonal error'. Note that, in practice, the sampling period is usually smaller than is suggested in figure 3.14, and an intersection closer to t is found.

In determining the orthogonal error, the visual background of this error introduces an extra complication: the incompatibility between *output* values and *time* values. For example, if figure 3.14 had been plotted on a different time scale, the picture would have been stretched horizontally. Other values for the orthogonal error would then have been found. This is illustrated by the triangle in figure 3.14, in which the orthogonal error is denoted ϵ. Because the dimension of the process and model outputs is different from that of the time, the orthogonal error should be calculated as:

$$\epsilon = \text{sign}(\Delta y)\sqrt{(\Delta y)^2 + \alpha(\Delta t)^2}$$

in which α is a weighting factor which weighs the relative importance of the time with respect to the output. A 'good' choice of α depends on the problem at hand and can best be determined in simulation. Among others, the best α depends on the process time constants and the model time constants. Relatively large time constants require a small weighting of the time influence. It appears that small values of α are generally favourable, and even a zero value may sometimes give the best results. Such a choice involves the use of Δy only as the error signal. Note that if the horizontal shift between y_p and y_m is completely due to the time delay and not to a difference in dynamics, Δt is nearly equivalent to the dead time. Δy in figure 3.14 is then the actual error between the process and model outputs, from which the effect of the dead time is almost completely removed.

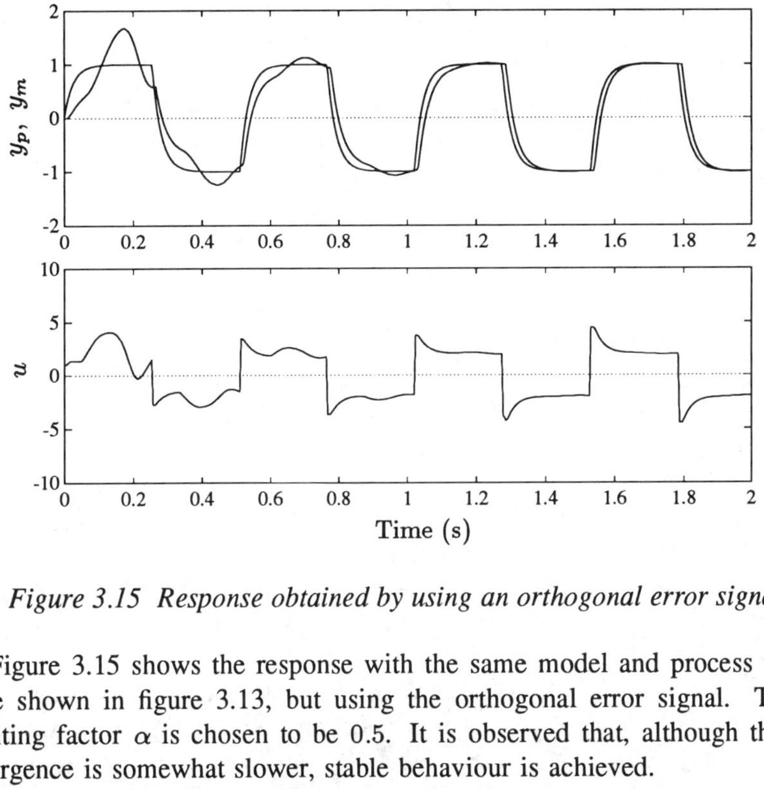

Figure 3.15 Response obtained by using an orthogonal error signal

Figure 3.15 shows the response with the same model and process transfers as are shown in figure 3.13, but using the orthogonal error signal. The time weighting factor α is chosen to be 0.5. It is observed that, although the initial convergence is somewhat slower, stable behaviour is achieved.

The above algorithm is difficult to analyze theoretically because of the complex error definition. A stability proof can therefore not be given, and simulations have shown that a large time delay still cannot be dealt with sufficiently by the algorithm. However, for a limited dead time magnitude, the use of an orthogonal error signal provides the required robustness. An advantage is that no knowledge of the magnitude of the dead time is needed, as opposed to many discrete-time schemes. The philosophy behind this modification is to make the error signal which is used in the adaptation insensitive to a known disturbance, which will also play an important role in the following chapters.

3.4 Summary

This chapter has discussed the subject of robustness of the adaptation in MRAC systems against external disturbances and a limited number of structural disturbances. External disturbances result in an output disturbance on the error equation. This disturbance can lead to unstable system behaviour if the degree of persistent excitation of the signal vector lies below a threshold, which is determined by the magnitude of the disturbance and the system characteristics.

If the PE condition cannot be guaranteed to be met, modifications of the adaptive law can be used to improve the robustness. Depending on the form of the modification, knowledge may be needed about an upper bound on the disturbance, or on the norm of the parameter vector. The modifications described in this section all have a simple form. The modification using a continuous adjustment of the adaptation pole, based on the output error magnitude (γ-modification), appears to give the best performance if measurement noise is present. The value of γ has a lower bound to guarantee stability in the presence of disturbances, and an upper bound to guarantee convergence if no disturbances are present. These bounds depend on the disturbance magnitude and the degree of PE of the signal vector respectively. Although simulations remain necessary to find a good value for γ, γ-modification is favoured over the others because it provides proper protection against disturbances over a wide range of the disturbance magnitude.

Process input offset, which is also an external disturbance, requires a special approach because a varying reference signal leads to nonconverging behaviour. An extra element '1' added to the signal vector, in combination with an extra adjustable parameter, gives good results. Extension to any deterministic input disturbance of which the form is known is straightforward.

The observation that during transients even a small dead time can result in a large error signal has resulted in the introduction of an 'orthogonal error signal'. For not too large dead time values, this gives a proper robustness improvement. The new error signal is defined such that the effect of the dead time is as well as possible removed, and hence the error is made insensitive to the disturbance. A stability proof cannot be given because of the complex error definition.

Problems

3.1 Assume you want to control a second-order process with states x_{p1} and x_{p2} and input u by state feedback, $u = k_0 r + k_1 x_{p1} + k_2 x_{p2}$. Assuming the perfect model matching is satisfied and no disturbances are present, do you expect parameter convergence for the following reference signals?

(a) a step: $r = 0$ for $t < 0$, $r = 1$ for $t \geq 0$

(b) a sine wave: $r = \sin(\omega t)$

(c) an addition of a step and a sine function: $r = 1 + \sin(\omega t)$

3.2 As stated in the section on γ-modification (page 95), for a given sufficient degree of PE of the reference signal, the adaptive system still converges to $\phi = 0$ for values of γ which are below a certain threshold (which depends on the degree of PE of the reference signal), that is, if no disturbances are present. Do you expect a similar result for σ modification?

3.3 The dead zone has some resemblance to σ-modification, in that parameter convergence in the absence of disturbances does not occur. Making σ-modification dependent on the error magnitude has led to γ-modification. Do you consider it possible to alter the dead zone principle similarly, such that its negative effects are relieved?

3.4 Consider a second-order system in which one of the states is disturbed by a disturbance δ:

$$\begin{aligned} \dot{x}_1 &= x_2 + \delta \\ \dot{x}_2 &= u \\ y &= x_1 \end{aligned}$$

Assume that δ is a ramp-like function:

$$\delta = A \cdot t$$

Hence, compensation of the disturbance is possible by adding a constant to the control u:

$$u = k_0 r + k_1 x_1 + k_2 x_2 + k_3$$

Explain why, as stated in section 3.3.2, 'process and model states may not be comparable due to the disturbance', and how this results in the inability to implement the adaptive compensation. Do you expect the same problem when using the augmented error method?

4

Discrete-Time MRAC

In the previous chapters, the design of model reference adaptive control systems was based on a continuous-time approach. The implementation of these algorithms, however, is mostly carried out on digital computers because of several limitations of analogue circuitry. If the computer sampling period is small compared to the system's time constants, analogue operators (integrators, etc.) can easily be approximated by their digital equivalents. However, a more formal approach is needed if this condition is not met.

In addition, the digital computer enables implementation of more complex algorithms than are possible in a continuous-time implementation. The development of these algorithms requires a discrete-time approach to the adaptive control problem, which is the subject of this chapter.

4.1 Introduction

Model reference adaptive control systems were originally developed using continuous-time analysis, stimulated by (among others) the presence of Lyapunov's and Popov's stability theories. It is therefore not surprising that the first steps towards discrete-time MRAC were based on the existing continuous-time designs. Landau (1979) derived adaptive laws for the discrete counterpart of the MRAC scheme that make use of all state variables of the process (section 2.1). However, because of the unclear link between the state variables of the continuous-time physical process and those of the discrete-time reference model, most discrete-time designs are developed for single-input, single-output processes. Ionescu and Monopoli (1977) and Narendra and Lin (1980) transformed the continuous-time augmented error method into a discrete form using Lyapunov's method. Several others (Suzuki and Takashima, 1978; Unbehauen, 1985 and 1987) derived the adaptive laws using hyperstability theory, each of them contributing several changes and enhancements to the design but leaving the main structure unaltered.

The development of these discrete-time counterparts of originally continuous-time designs requires discrete forms of existing stability theories. Lyapunov's stability theory and Popov's hyperstability theory both have a discrete counterpart that can be used in the design of discrete-time MRAC. The inherent one-step delay in the adaptation in the discrete case needs a correction factor for the output error, linking the *a priori* to the *a posteriori* error (see section 4.2).

Further, speaking in terms of hyperstability theory, using new adaptation schemes based on least-squares estimation methods violates the passivity property of the nonlinear part of the error equation. Landau and Silveira (1979) and Landau (1980) modified Popov's hyperstability theory in such a way that requirements on the linear part can be exchanged with those on the nonlinear part, which was made intuitively clear in chapter 1. This means that the SPR requirement on the linear part may be violated, provided that the nonlinear part satisfies a heavier passivity demand than originally, and vice versa. An example of this has already been given in section 1.3.4, in which proportional adaptation allowed a non-SPR error equation. The availability of the modified stability method enables a smooth introduction of more complex adaptation schemes (such as least-squares methods), maintaining the stability proof.

Later, MRAC schemes were introduced which were completely discrete-time oriented, and did not have a continuous-time predecessor. The design of these newer forms of discrete MRAC starts with the assumption that all process parameters are known. A primary controller is then designed that makes the transfer of the controlled process equal to that of the reference model. Finally, the controller is extended by a parameter estimation scheme (usually based on a least-squares method) that adjusts the controller parameters. By analyzing the parameter estimation part and the control part simultaneously, stable adaptive laws can be obtained.

This chapter gradually introduces discrete-time model reference adaptive control, starting with some specific differences from the continuous-time approach. Using the discrete form of the augmented error method, simulation examples illustrate the application of discrete MRAC. This method is essentially based on the continuous-time version, but can be expanded simply with, for example, a least-squares adaptation method. To give more insight into the stability properties of adaptive systems using least-squares adaptation, the hyperstability theory is expanded such that it can be applied to such systems. This paves the way for the introduction of newer discrete-time model reference adaptive schemes that do not have a continuous-time predecessor.

4.2 Going from continuous to discrete MRAC

As in the continuous case, in discrete-time MRAC design stability theory plays a vital role. Both Lyapunov's method and Popov's hyperstability theory can be extended to the discrete domain. This section describes both methods and shows an important aspect in discrete MRAC: the presence of a one-step delay in the adaptation loop.

4.2.1 Lyapunov's method

In order to establish stability, a Lyapunov function $V(\boldsymbol{x})$ must be chosen as a function of the (discrete) system states. The requirements are similar to those in the continuous case (see equation (1.2)), with an obvious change from $V(t)$ to $V(k)$. To guarantee asymptotic stability, $\Delta V(k) = V(k) - V(k-1)$ must be negative definite. To apply Lyapunov's method, the error equations must first be established. Next, a suitable Lyapunov function must be chosen, which is usually a quadratic function of both the state error and the parameter error. The demand $\Delta V(k) < 0$ then leads to stable adaptive laws, which have the well-known integral form. A difference with regard to the continuous-time case arises in deriving the compensator $\boldsymbol{P}$ for the linear part. Instead of using the standard Lyapunov equation (1.3), if the matrix $\boldsymbol{A}$ of the system $\boldsymbol{x}(k) = \boldsymbol{A}\boldsymbol{x}(k-1)$ is asymptotically stable, a positive definite matrix $\boldsymbol{Q}$ yields a positive definite matrix $\boldsymbol{P}$ according to the discrete-time Lyapunov equation:

$\boldsymbol{A}^T\boldsymbol{P}\boldsymbol{A} - \boldsymbol{P} = -\boldsymbol{Q}$

This equation can easily be derived by choosing a Lyapunov function $V(\boldsymbol{x}) = \boldsymbol{x}^T\boldsymbol{P}\boldsymbol{x}$ and calculating ΔV:

$$\begin{aligned}
V(k) &= \boldsymbol{x}^T(k)\boldsymbol{P}\boldsymbol{x}(k) \\
V(k-1) &= \boldsymbol{x}^T(k-1)\boldsymbol{P}\boldsymbol{x}(k-1) \\
\Delta V(k) &= V(k) - V(k-1) \\
&= \boldsymbol{x}^T(k-1)\boldsymbol{A}^T\boldsymbol{P}\boldsymbol{A}\boldsymbol{x}(k-1) - \boldsymbol{x}^T(k-1)\boldsymbol{P}\boldsymbol{x}(k-1) \\
&= \boldsymbol{x}^T(k-1)\left(\boldsymbol{A}^T\boldsymbol{P}\boldsymbol{A} - \boldsymbol{P}\right)\boldsymbol{x}(k-1)
\end{aligned}$$

For an asymptotically stable system ΔV is negative definite, and so:

$$\begin{aligned}
\Delta V &= -\boldsymbol{x}^T(k-1)\boldsymbol{Q}\boldsymbol{x}(k-1) \\
\Longrightarrow \boldsymbol{A}^T\boldsymbol{P}\boldsymbol{A} - \boldsymbol{P} &= -\boldsymbol{Q}
\end{aligned}$$

4.2.2 Hyperstability

Popov's hyperstability theory (section 1.3.4) also has a discrete counterpart. In order to meet the SPR requirement, a similar condition to that in the continuous case can be formulated for the linear part (Landau, 1979), which is called the Kalman–Szegö–Popov lemma. This lemma is, however, difficult to apply in practice, and it is therefore more usual to transform the discrete transfer function to the s-domain, and carry out the analysis there. The demand on the nonlinear part is the discrete Popov passivity demand (compare equation (1.6)):

$$\sum_{k=0}^{k_1} \epsilon(k)w(k) \geq -\chi, \quad \forall\, k_1 \geq 0, \quad \chi > 0 \tag{4.1}$$

Here, $\epsilon(k)$ is the error signal to be used in the adaptation and $w(k)$ is the output of the nonlinear part.

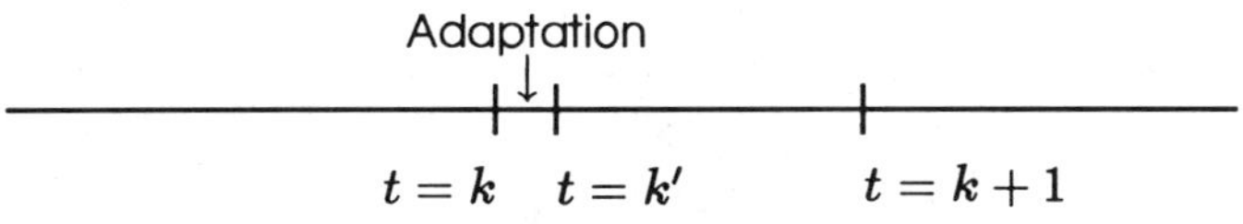

Figure 4.1 One-sample delay in the adaptation, present in discrete adaptive control

4.2.3 One-sample delay

In discrete time, the effect of a change in the control parameter values will be noticed only one sample period later, as shown in figure 4.1. At $t = k$ the process output is measured and transferred to the computer. The adaptation starts and at $t = k'$ the controller parameters are adjusted and sent to the primary controller. Although the effect of the parameter change starts at $t = k'$, the new process output is not measured until $t = k + 1$. Note that in discrete systems a parameter change is likely to have a *direct* influence on the output, as opposed to continuous-time systems. Now, for the direct application of stability methods the *a posteriori* process output $y_p(k)$ should be available, while these methods require it to be used in the adaptive law:

$$y_p(k) = W_c\,[\boldsymbol{\theta}(k)]\,u(k)$$

$W_c[\boldsymbol{\theta}(k)]$ represents the transfer function of the closed-loop system with the already adapted parameter vector $\boldsymbol{\theta}(k)$. Because the effect of the adaptation is noticed first at $t = k + 1$, only the *a priori* process output $y_p^\circ(k)$ is available, which depends on the parameters $\boldsymbol{\theta}(k-1)$ adjusted one sample period earlier:

$$y_p^\circ(k) = W_c\,[\boldsymbol{\theta}(k-1)]\,u(k)$$

In analogy with the above, an *a priori* output error $e_1^\circ(k) = y_p^\circ(k) - y_m(k)$ and an *a posteriori* output error $e_1(k) = y_p(k) - y_m(k)$ are defined. Calculating $e_1(k)$ from $e_1^\circ(k)$ needs a correction factor, which can easily be illustrated by a simple example.

Example 4.1 ***A priori / a posteriori*** **correction**

Assume an error equation of the form:

$$\epsilon(k) = y_p(k) - y_m(k) = \phi^T(k)\omega(k) \tag{4.2}$$

in which, for simplicity, the linear part is assumed to be 1. Such an error equation is not unusual in discrete MRAC systems. The parameters are updated using the adaptive law:

$$\theta(k) = \theta(k-1) - \Gamma\epsilon(k)\omega(k) \tag{4.3}$$

Because the linear part, consisting of a transfer of 1, is SPR, it remains to be shown that the nonlinear transfer from $\epsilon(k)$ to $w(k) = -\phi^T(k)\omega(k)$ satisfies Popov's inequality (4.1) in order to guarantee stability. To test this inequality, the update law (4.3) is written as:

$$\begin{aligned} \phi(k) &= \phi(k-1) - \Gamma\epsilon(k)\omega(k) \\ &= \phi(0) - \Gamma\sum_{i=1}^{k}\epsilon(i)\omega(i) \end{aligned} \tag{4.4}$$

Substituting $\phi(k)$, $\omega(k)$ and $\epsilon(k)$ in equation (4.1) yields:

$$\begin{aligned} \sum_{k=0}^{k_1}\epsilon(k)w(k) &= -\sum_{k=0}^{k_1}\epsilon(k)\phi^T(k)\omega(k) \\ &= -\sum_{k=0}^{k_1}\epsilon(k)\omega^T(k)\left\{\phi(0) - \Gamma\sum_{i=1}^{k}\epsilon(i)\omega(i)\right\} \end{aligned}$$

See notes pg 90 - 95

Similarly to the continuous-time case (examples 1.2 and 2.3), this expression is always larger than a term proportional to $-\frac{1}{2}\phi^T(0)\phi(0)$, and hence the adaptive law (4.3) satisfies Popov's criterion. The adaptive system is hyperstable. Unfortunately, the *a posteriori* output $y_p(k)$ in equation (4.2) is only available after the parameter adjustment at $t = k$, for which $\epsilon(k)$ and thus $y_p(k)$ must be known. Hence, for the calculation of $\theta(k)$, $\epsilon(k)$ is required, and for the calculation of $\epsilon(k)$, $\theta(k)$ is required. To solve this dilemma, $\epsilon(k)$ can be expressed in terms of the *a priori* error $\epsilon^\circ(k)$, by considering $\epsilon(k)$ and $\epsilon^\circ(k)$:

$$\begin{aligned} \epsilon(k) &= \phi^T(k)\omega(k) \\ \epsilon^\circ(k) &= \phi^T(k-1)\omega(k) \end{aligned}$$

By considering the adaptive law (4.3) and (4.4), the *a posteriori* error can be written as:

$$\epsilon(k) = \phi^T(k)\omega(k) = \phi^T(k-1)\omega(k) - \omega^T(k)\Gamma\epsilon(k)\omega(k)$$

And hence:

$$\begin{aligned} \left\{1 + \left[\omega^T(k)\Gamma\omega(k)\right]\right\}\epsilon(k) &= \epsilon^\circ(k) \\ \Longrightarrow \epsilon(k) &= \frac{\epsilon^\circ(k)}{1 + [\omega^T(k)\Gamma\omega(k)]} \end{aligned} \tag{4.5}$$

Equation (4.5) is known as the *a priori / a posteriori correction*, because it allows the use of the *a priori* error in the adaptation mechanism.

□

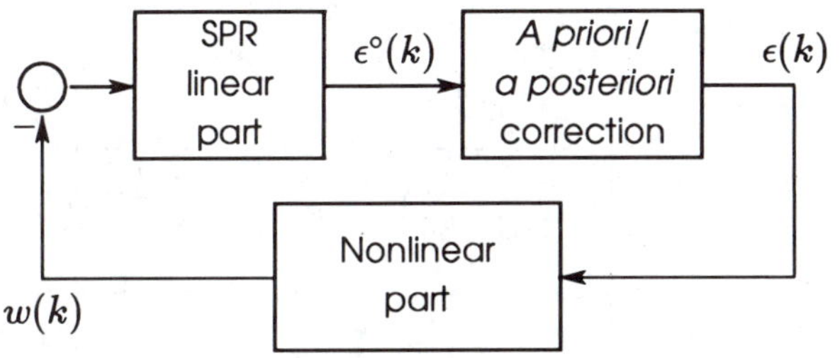

Figure 4.2 General error scheme for discrete MRAC, consisting of a linear feed-forward part and a nonlinear feedback part

A general error scheme looks like that shown in figure 4.2. Note that when the adaptation changes the parameters slowly (which is the case if Γ is chosen small or when the current signal vector $\omega(k)$ is small), the correction factor is small. However, if the adaptation has large effects, the error signal $\epsilon^\circ(k)$ is made smaller before being used in the adaptation mechanism. This modification is necessary for maintaining stability and compensates for the phase shift introduced by the one-sample delay. The form of the *a priori / a posteriori* correction resembles the normalization which is common in continuous-time MRAC (section 1.3.4).

4.3 The discrete-time augmented error method

The continuous-time augmented error method can be transformed into a discrete version without much difficulty (Narendra and Lin, 1980). The primary controller structure remains essentially unaltered and is shown in figure 4.3. The auxiliary

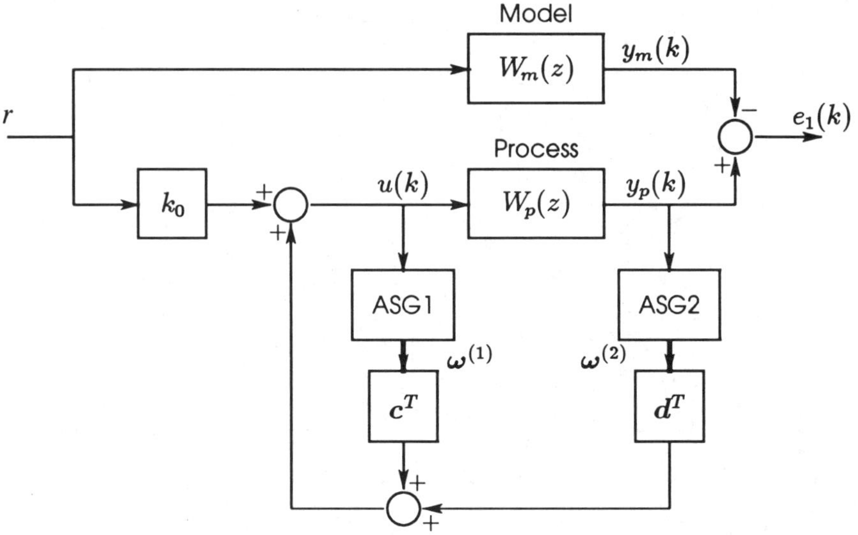

Figure 4.3 Controller scheme of the discrete-time augmented error method

signal generators ASG1 and ASG2 generate signal vectors $\boldsymbol{\omega}^{(1)}$ and $\boldsymbol{\omega}^{(2)}$, which are fed back to the process input through parameter vectors $\boldsymbol{c}$ and $\boldsymbol{d}$. There are two main differences between the discrete-time and the continuous-time augmented error method:

- The order of the auxiliary signal generators is increased to the process order n. Originally, this order was $n-1$. As a result of this increase, the perfect model-matching condition is satisfied if $m \leq n$ (m is the number of zeros of $W_p(z)$, n is the number of poles). In the continuous case, this condition is: $m \leq n-1$. Because the transfer function of a process for which $m \leq n-1$ in the continuous case also has $m \leq n-1$ in the z-domain, the original condition is more strict than necessary. Its relaxation leads to a possible simplification of the ASGs, as will be shown in section 4.4.

- The signal vector and parameter vector become:

$$\begin{aligned} \boldsymbol{\omega}^T &= \left(r,\ \boldsymbol{\omega}^{(1)T},\ \boldsymbol{\omega}^{(2)T}\right) \\ \boldsymbol{\theta}^T &= \left(k_0,\ \boldsymbol{c}^T,\ \boldsymbol{d}^T\right) \end{aligned}$$

  The signal vector contains one element more than in the continuous case ($2n+1$ total), but the process output $y_p(k)$ is excluded.

As in the continuous-time augmented error method, the zeros of the reference model transfer function must be present in the denominator of the auxiliary signal generators. In the continuous case, zeros are not always present. Process transfer functions described in z usually have zeros, and so the choice of the ASGs is subject to more restrictions than in the continuous case.

Lyapunov's stability method is used in deriving the adaptive laws. An error-augmenting signal is added to the output error to obtain an SPR transfer function $W_m L$ of the linear part:

$$\begin{aligned} e_1(k) &= y_p(k) - y_m(k) \\ \epsilon(k) &= e_1(k) - W_m L(z)\left[L^{-1}(z)\boldsymbol{\theta}^T(k) - \boldsymbol{\theta}^T(k)L^{-1}(z)\right]\boldsymbol{\omega}(k) \end{aligned}$$

The filters L^{-1} are chosen such that $W_m L$ is SPR. A simple choice is: $L^{-1} = W_m$, making $W_m L = 1$. However, choosing L^{-1} different from W_m can improve the system behaviour, as will be shown in the next section. The following parameter update law is used:

$$\boldsymbol{\theta}(k) = \boldsymbol{\theta}(k-1) - \boldsymbol{\Gamma}\epsilon(k)\boldsymbol{\xi}(k), \quad \text{with: } \boldsymbol{\xi}(k) = L^{-1}\boldsymbol{\omega}(k)$$

Now, only the *a priori* output error $e_1^\circ(k) = y_p^\circ(k) - y_m(k)$, and the *a priori* augmented error ϵ° are available. In other words, calculation of $\boldsymbol{\theta}(k)$ is only possible if $\epsilon(k)$ is available, which in turn depends on $\boldsymbol{\theta}(k)$ in the error-augmenting part. To calculate $\boldsymbol{\theta}(k)$ from the measured *a priori* error, an *a priori / a posteriori* correction is required:

$$\boldsymbol{\theta}(k) = \boldsymbol{\theta}(k-1) - \boldsymbol{\Gamma}\frac{\epsilon^\circ(k)\boldsymbol{\xi}(k)}{1 + W_m L\left[\boldsymbol{\xi}^T(k)\boldsymbol{\Gamma}\boldsymbol{\xi}(k)\right]} \tag{4.6}$$

Compared to equation (4.5) (in which the linear part of the error equation was assumed to be 1), $W_m L$ appears in the denominator (the new linear part of the error equation), and $\boldsymbol{\omega}$ has been replaced by $\boldsymbol{\xi}$ (the new signal vector, used in the adaptive law). The derivation of equation (4.6) is similar to that given in section 4.2.3. In Narendra and Lin, 1980; Lin and Narendra, 1980, the same result is found by making use of a new error model.

4.4 An example using discrete MRAC

The discrete augmented error method is tested in simulation with the earlier-used process and model transfer functions (the sampling period is 0.1 s):

$$W_p(s) = \frac{8}{s^2+2s+8}$$

$$W_m(s) = \frac{16}{s^2+8s+16} \Longrightarrow W_m(z) = 0.109\,\frac{z}{z^2-1.341z+0.449}$$

$W_m(z)$ has one zero at $z=0$ and two poles at $z=0.67$. The filters L^{-1} are chosen equal to the reference model transfer W_m. This implies that $W_mL=1$ and so:

$$\begin{aligned}\epsilon(k) &= e_1(k) - \left[W_m\boldsymbol{\theta}^T(k)\boldsymbol{\omega}(k) - \boldsymbol{\theta}^T(k)W_m\boldsymbol{\omega}(k)\right]\\ &= e_1(k) - \left[W_m u(k) - \boldsymbol{\theta}^T(k)\boldsymbol{\xi}(k)\right]\end{aligned}$$

4.4.1 The auxiliary signal generators

The form of the auxiliary signal generators is as shown in figure 4.4, in which ASG1 in combination with the adjustable parameters c_1 and c_2 is shown. The transfer function is:

$$W_{\text{ASG1}} = \frac{c_2z^{-1}+c_1z^{-2}}{1+n_1z^{-1}+n_2z^{-2}} = \frac{c_2z+c_1}{z^2+n_1z+n_2}$$

The denominator polynomials of the ASGs must contain the zeros of the reference model, and because the order is equal to the reference model order ($n=2$) while only one process zero is present, one pole can be chosen freely. The placement of this pole is important: both a too fast ASG (pole close to 0 in the z-plane)

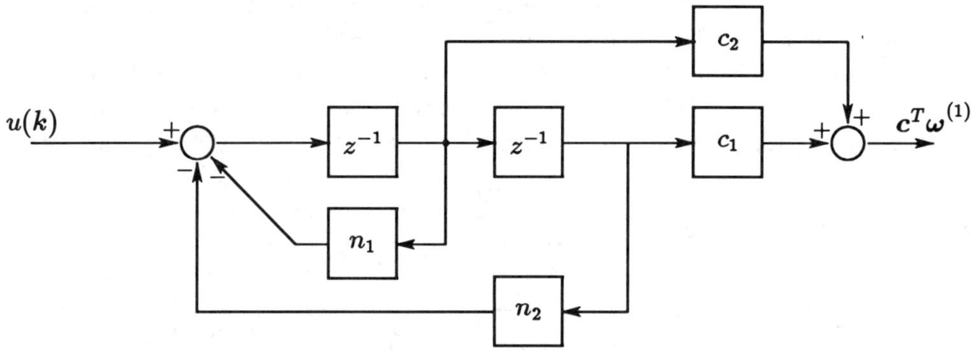

Figure 4.4 Auxiliary signal generator 1, *generating* $\omega^{(1)}$

and a too slow ASG (pole close to 1) degrades the convergence properties. As a rule, the pole is best placed such that the ASG bandwidth resembles that of the reference model. Figure 4.5 shows the system response if the ASG pole is placed at $z = -0.77$, which appears to be a good compromise. Note that the other ASG pole is placed at $z = 0$, at the same place as the reference model zero.

The discrete augmented error structure is based on the ability to control processes for which $m \leq n$. However, a condition $m \leq n - 1$ is sufficient

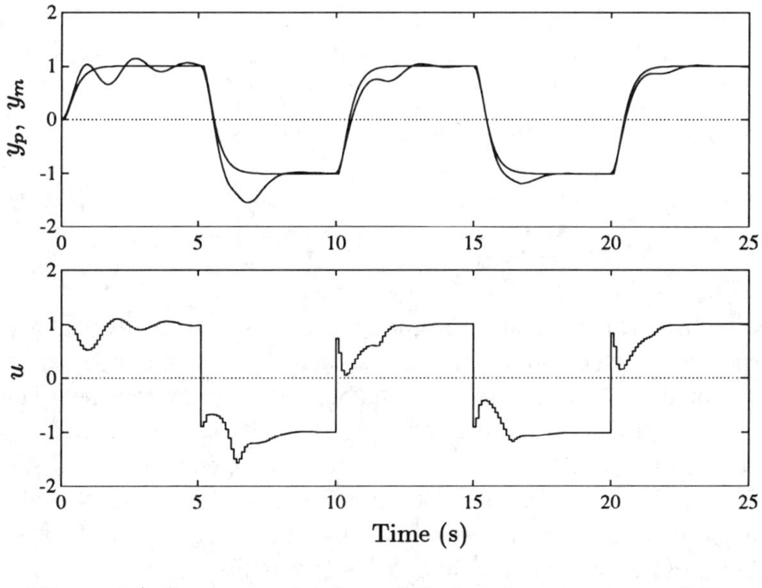

Figure 4.5 Response with 'normal' signal generators and filters

because the z-transform of a continuous-time process always has $m \leq n-1$. The poles of the ASGs appear as zeros in the closed-loop transfer function, and therefore the ASG order must at least be equal to m, which maximally equals $n-1$. Hence, the ASG orders can be reduced from n to $n-1$, as shown in figure 4.6. Writing the new ASG transfers, including the adjustable parameters,

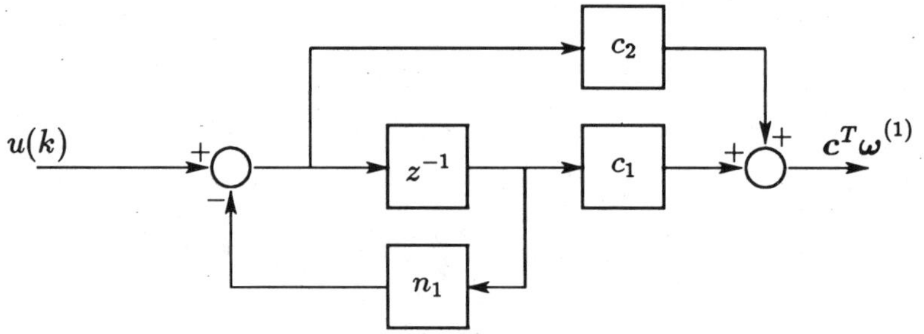

Figure 4.6 Block diagram of reduced auxiliary signal generator 1

as:

$$
\begin{aligned}
\text{ASG1: } \frac{C}{N}, \text{ with } C &= c_n z^{n-1} + \ldots + c_2 z + c_1 \\
N &= z^{n-1} + n_{n-1} z^{n-2} + \ldots + n_1 \\
\text{ASG2: } \frac{D}{N}, \text{ with } D &= d_n z^{n-1} + \ldots + d_2 z + d_1
\end{aligned}
\tag{4.7}
$$

the closed-loop transfer function of figure 4.3 becomes:

$$
\frac{y_p}{r} = \frac{k_0 k_p Z_p N}{(N-C)R_p - k_p D Z_p}
\tag{4.8}
$$

In equation (4.8), Z_p and R_p are the process numerator and denominator polynomials, respectively, which are monic in z. k_p is the gain of the process, using the same representation of W_p as in the continuous case. By an argument similar to that given in section 2.5, this structure can be shown to satisfy the perfect model-matching condition. Note that the total number of parameters has not changed due to the ASG simplification. Because the modified ASGs allow satisfaction of the perfect model-matching requirement, the modification is allowed without affecting the stability proof. An algebraic loop arises for ASG1 which can, however, be solved analytically.

Simulations have shown that reducing the ASG order produces similar results as in the case of full-order ASGs where the position of the extra pole is chosen with care. The main advantage of reducing the ASGs therefore lies in relieving the designer of the task of choosing a proper pole placement for the ASGs.

4.4.2 The filters L^{-1}

The filters L^{-1} are used in the error augmentation, and generate the signal vector $\boldsymbol{\xi} = L^{-1}\boldsymbol{\omega}$ used in the adaptation. For stability, $W_m L$ must be SPR and, for simplicity, L^{-1} is chosen equal to W_m. This is of course an arbitrary choice, and it can be advantageous to choose L^{-1} different from W_m. This is shown in

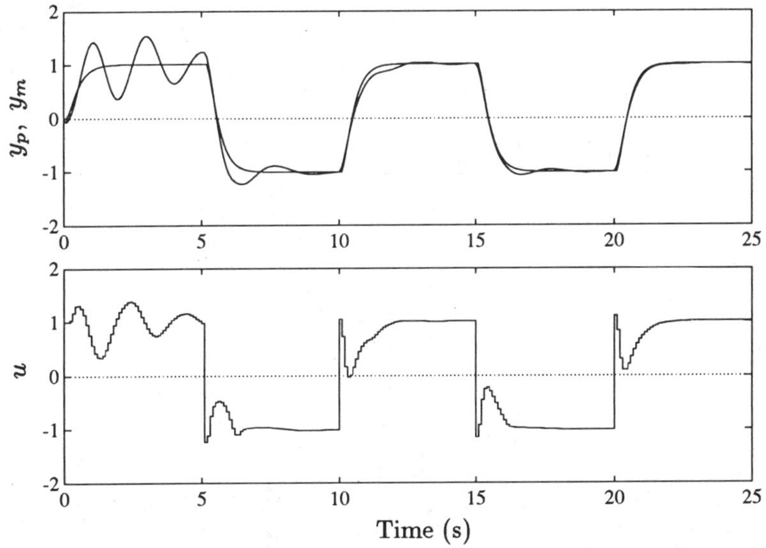

Figure 4.7 Response obtained by using fast filters L^{-1}

figure 4.7, in which the bandwidth of L^{-1} is twice that of W_m:

$$L^{-1} = \frac{0.301z}{z^2 - 0.899z + 0.202}$$

This choice still guarantees an SPR error equation:

$$W_m L = 0.358\,\frac{z^2 - 0.899z + 0.202}{z^2 - 1.341z + 0.449}$$

As shown in figure 4.7, convergence is improved considerably, which corresponds to the results in section 2.7.

4.4.3 Least-squares adaptation

Section 4.5 will introduce the least-squares adaptation method as an alternative to gradient adaptation. This section presents results with this type of adaptation mechanism applied in the augmented error method (figure 4.8). Here, the forget-

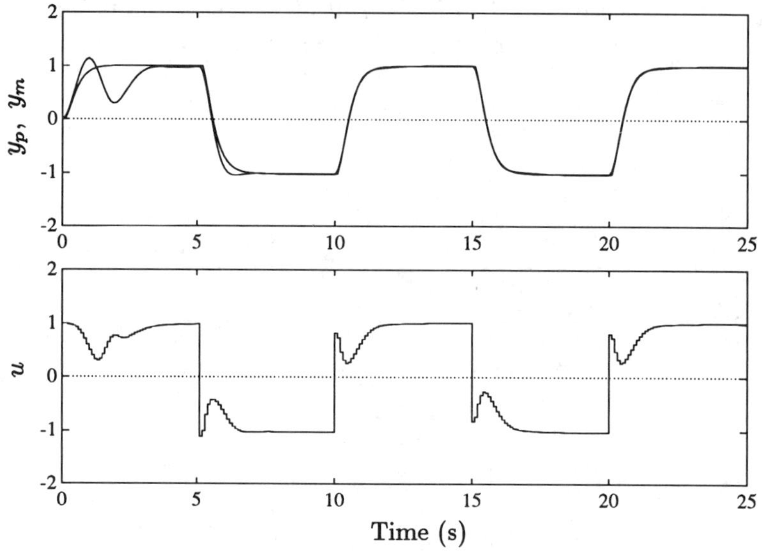

Figure 4.8 Response obtained by using least-squares adaptation

ting factor $\lambda = 0.95$. Observe that the least-squares adaptation provides a faster adaptation than the original gradient adaptation. The filters L^{-1} are chosen equal to W_m and so the linear part of the error equation becomes 1. Increasing the filter's bandwidth does not appear to improve the behaviour. This will be confirmed theoretically in section 4.5, where it will be shown that for least-squares adaptation an SPR error equation is no longer sufficient, and thus the allowed filter choice is limited more strictly.

4.5 Least-squares adaptation methods

If the standard gradient-type adaptation algorithm is used, the passivity demand on the nonlinear part of the error equation is met, and stability is established by making the linear part SPR. A correction for the *a priori/a posteriori* error completes the design. However, if an adaptation mechanism is applied that uses time-varying adaptation gains, such as least-squares adaptation, the passivity demand on the nonlinear part is generally not satisfied (Landau and Silveira, 1979; Landau, 1980). A modification of stability theory is required to allow for such a violation.

4.5.1 A modification of hyperstability theory

In order to be able to obtain a stability proof for more types of adaptation, the hyperstability theory has to be extended. For this extension, two 'system classes' are defined (Landau and Silveira, 1979; Landau, 1980). The first class is called $L(\boldsymbol{\Lambda})$ and applies to linear, time-invariant discrete systems:

$$\begin{aligned} \boldsymbol{x}(k+1) &= \boldsymbol{A}\boldsymbol{x}(k) + \boldsymbol{B}\boldsymbol{u}(k) \\ \boldsymbol{y}(k) &= \boldsymbol{C}\boldsymbol{x}(k) + \boldsymbol{D}\boldsymbol{u}(k) \end{aligned} \tag{4.9}$$

This system belongs to the class $L(\boldsymbol{\Lambda})$ if a new system, resulting from a parallel configuration of equation (4.9) and a gain matrix $-\frac{1}{2}\boldsymbol{\Lambda}$, is SPR (figure 4.9). The use of a gain factor $\frac{1}{2}$ will become clear later in this section. A system H which belongs to a class $L(\boldsymbol{\Lambda})$, with $\boldsymbol{\Lambda} > 0$, is 'super'-SPR, which means that the Nyquist diagram lies to the right of a positive constant c for all ω: $\mathrm{Re}[H(j\omega)] > c$. This implies that the relative order of H must be zero to have $|H(j\omega)| \neq 0$ for $\omega \to \infty$. This can be shown by a simple example.

Example 4.2 A super-SPR transfer function

Consider a discrete-time system with transfer function:

$$H(z) = 0.2\,\frac{1 - 0.5z^{-1}}{1 - 0.9z^{-1}}$$

The Nyquist diagram of $H(z)$ is shown in figure 4.10, curve (a). Because the

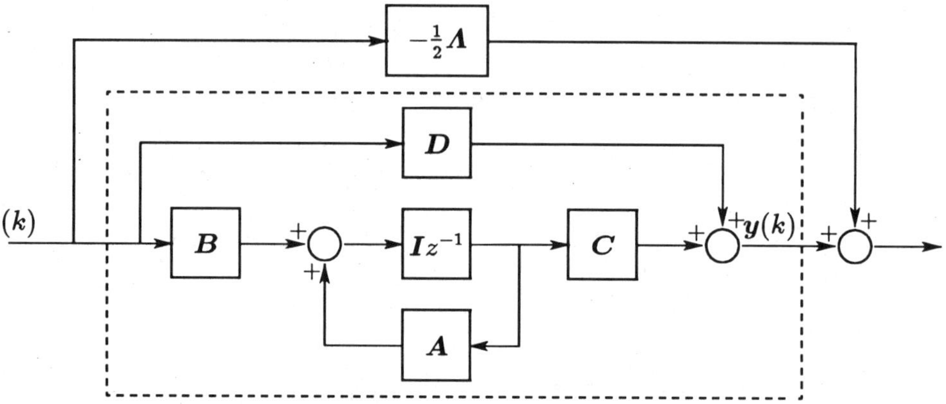

Figure 4.9 Parallel linear system used for extending hyperstability theory

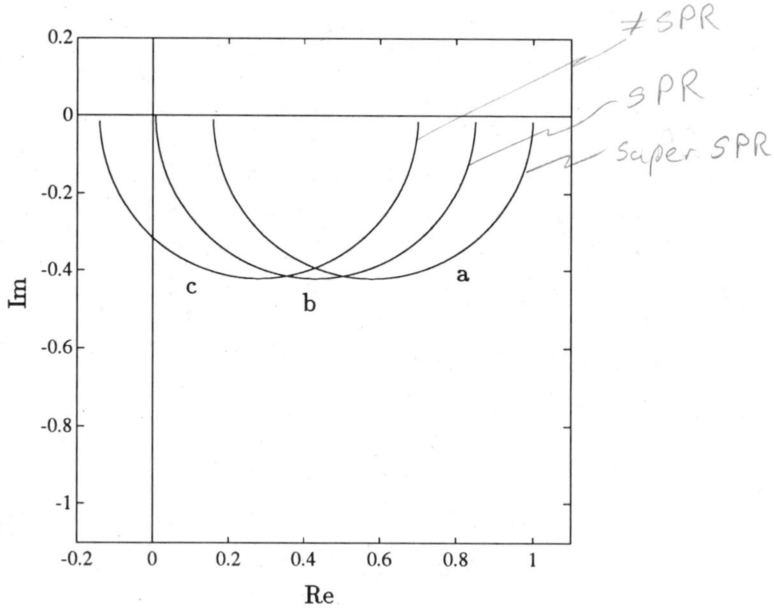

Figure 4.10 Example 4.2, Nyquist diagrams of $H(z) - \frac{1}{2}\lambda$*:* (a) $\lambda = 0$; (b) $\lambda = 0.30$*; and* (c) $\lambda = 0.60$

Nyquist diagram lies completely in the right-half plane, this transfer function is SPR. Now, by subtracting $\frac{1}{2}\lambda = 0.15$ and $\frac{1}{2}\lambda = 0.30$, respectively, Nyquist curves (b) and (c) result. Observe that for $\frac{1}{2}\lambda = 0.15$ the system is on the edge of losing its SPR property, and for larger values of λ the parallel system loses its SPR property. $H(z)$ therefore belongs to the class $L(0.3)$. The higher λ is allowed to grow before the system loses its SPR property, the more 'super'-SPR the system is. Note that a transfer $H(z) = 1$ belongs to a class $L(2)$. □

The second system class is called $N(\boldsymbol{\Pi})$ and is used for the description of nonlinear, time-varying systems. Such a system is said to belong to the class $N(\boldsymbol{\Pi})$ if the original time-varying system, to which a feedback loop with gain $\frac{1}{2}\boldsymbol{\Pi}$ is added, satisfies Popov's inequality (equation (4.1)). The resulting system is shown in figure 4.11. The feedback $\frac{1}{2}\boldsymbol{\Pi}$ can make a nonpassive system passive ($\boldsymbol{\Pi}$ can

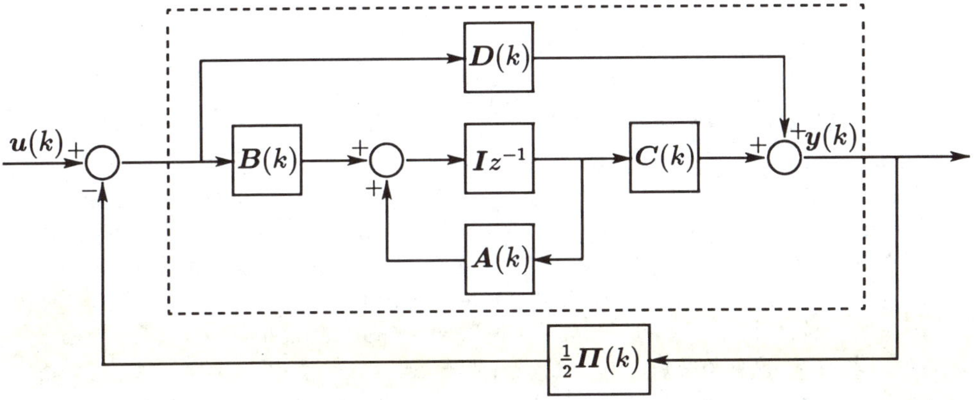

Figure 4.11 Nonlinear, time-varying feedback part

in itself be time varying). The 'amount of feedback' $\frac{1}{2}\boldsymbol{\Pi}$ that is needed to make the system passive determines the 'class' that the system belongs to. The larger $\boldsymbol{\Pi}$ must be, the further the system is from satisfying Popov's inequality.

Now, having defined the two classes, it can be proven that a system, consisting of a linear subsystem of class $L(\boldsymbol{\Lambda})$ in the feedforward path and a time-varying system belonging to class $N(\boldsymbol{\Pi})$ in the feedback path, is asymptotically stable if (Landau and Silveira, 1979; Landau, 1980):

$$\boldsymbol{\Lambda} \geq \boldsymbol{\Pi}$$

This means in practice that a deviation of the nonlinear part from Popov's inequality is allowed, provided that the linear part is 'super'-SPR. The above theorem can

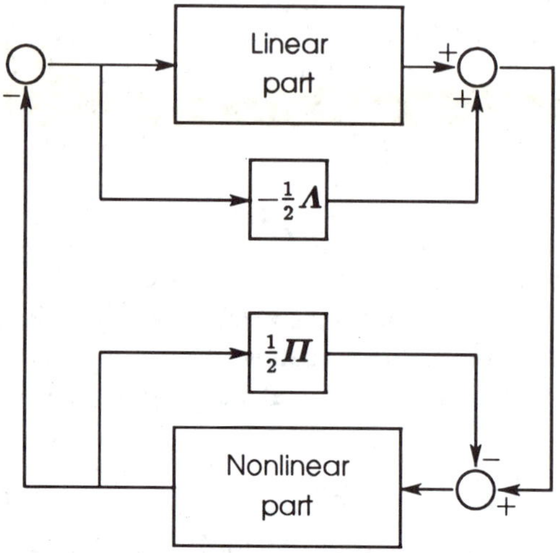

Figure 4.12 Feedback configuration of linear and nonlinear parts

be used to determine to which class of $N(\boldsymbol{\Pi})$ a specific adaptation mechanism belongs. The requirement on the linear part can then be determined.

Note that adding a feedforward $-\frac{1}{2}\boldsymbol{\Lambda}$ to the linear system and a feedback $\frac{1}{2}\boldsymbol{\Pi}$ to the nonlinear part leads to the configuration shown in figure 4.12. If the linear part with feedforward $-\frac{1}{2}\boldsymbol{\Lambda}$ is SPR, and the nonlinear part with feedback $\frac{1}{2}\boldsymbol{\Pi}$ is passive, the overall system is stable according to the hyperstability theory. Now, if $\boldsymbol{\Lambda} = \boldsymbol{\Pi}$, the feedforward $-\frac{1}{2}\boldsymbol{\Lambda}$ and the feedback $\frac{1}{2}\boldsymbol{\Pi}$ cancel each other, and hence figure 4.12 becomes equal to the original feedback without $\boldsymbol{\Lambda}$ and $\boldsymbol{\Pi}$. Therefore, if the structure of figure 4.12 is stable, then so is the feedback configuration of the linear and the nonlinear parts. This agrees with the above result.

4.5.2 Least-squares parameter updating

Until now, the parameters were updated with a fixed adaptation gain $\boldsymbol{\Gamma}$. In applying a least-squares-type algorithm, this gain is recursively computed during each iteration step using the recursion formula:

$$\boldsymbol{\Gamma}(k+1) = \frac{1}{\lambda_1}\left[\boldsymbol{\Gamma}(k) - \frac{\boldsymbol{\Gamma}(k)\boldsymbol{\omega}(k-d)\boldsymbol{\omega}^T(k-d)\boldsymbol{\Gamma}(k)}{(\lambda_1/\lambda_2) + \boldsymbol{\omega}^T(k-d)\boldsymbol{\Gamma}(k)\boldsymbol{\omega}(k-d)}\right] \tag{4.10}$$

where: $0 < \lambda_1 \leq 1, \quad 0 \leq \lambda_2 < 2; \quad \boldsymbol{\Gamma}(0) > 0$

In equation (4.10), $\boldsymbol{\omega}(k)$ is the system's signal vector, and d is the process time delay. The parameter λ_1 (or, actually, (λ_1/λ_2)) corresponds to the forgetting factor in least-squares estimation, and λ_2 is included mainly for generalization purposes: by putting $\lambda_1 = 1$ and $\lambda_2 = 0$, the original fixed-gain adaptation appears. If $\lambda_2 = 1$ equation (4.10) becomes equal to the usual least-squares formula. Due to the time-varying adaptation gain $\boldsymbol{\Gamma}$ the nonlinear part is no longer passive, but can be shown to belong to the class $N(\lambda_2)$ (Landau, 1980).

The above recursion formula, in combination with a parameter update mechanism:

$$\boldsymbol{\theta}(k) = \boldsymbol{\theta}(k-1) + \boldsymbol{\Gamma}(k)\boldsymbol{\omega}(k-d)\epsilon(k) \tag{4.11}$$

results in an asymptotically stable error signal $\epsilon(k)$, provided $\epsilon(k)$ is generated by:

$$\epsilon(k) = H(z)\left[\boldsymbol{\phi}^T(k)\boldsymbol{\omega}(k-d)\right]$$

In this equation, $\boldsymbol{\phi}(k) = \boldsymbol{\theta}^* - \boldsymbol{\theta}(k)$, which has a sign difference from the continuous-time definition. This explains the '+' sign in the parameter update law (equation (4.11)). The linear part of the error transfer function, $H(z)$, must be 'super'-SPR because the adaptation mechanism belongs to the class $N(\lambda_2)$. If $\lambda_2 = 1$, stability is guaranteed if $H(z) - \frac{1}{2}$ is SPR. Note that gradient adaptation ($\lambda_2 = 0$) belongs to the class $N(0)$ and therefore the original SPR requirement reappears.

The introduction of least-squares adaptation in effect means the use of a time-varying adaptation gain matrix. In addition, the elements outside the diagonal of this matrix can become nonzero, which implies that the increment of a parameter not only depends on the corresponding element in the signal vector, but also on other elements. Simulations have shown that the inclusion of these elements may not be omitted.

In practice, implementation of the above least-squares scheme is dangerous if $\lambda_1 < 1$, because the matrix $\boldsymbol{\Gamma}$ may become unbounded if the degree of PE of the signal vector is low. This is a common problem of the least-squares method. Measures have to be taken to prevent $\boldsymbol{\Gamma}(k)$ from becoming too large. To achieve this, several modifications are available, such as limiting or scaling $\boldsymbol{\Gamma}(k)$. A simple and effective solution is limiting the trace of $\boldsymbol{\Gamma}(k)$. If the trace is smaller than some prespecified value $\boldsymbol{\Gamma}(k)$ is unaltered, but if it is larger $\boldsymbol{\Gamma}(k)$ is scaled down such that the trace is limited to its maximum value.

4.6 Independent tracking and regulation

The discrete-time augmented error method, as described in the sections 4.3 and 4.4, evolved from its continuous-time predecessor. Because dealing with polynomials in z is not much different from dealing with their s-equivalents, this did not impose many problems other than typically discrete-time effects. A main difference was the use of an *a priori / a posteriori* correction, required to implement the adaptive law using known signals. The primary controller structure remained essentially unchanged.

In the early 1980s, Landau and Lozano (1981) and Lozano and Landau (1981) presented a new MRAC scheme which made it possible to specify the desired tracking and regulation properties separately, whereas existing MRAC schemes only dealt with tracking requirements. The new scheme is called 'independent tracking and regulation' (ITR). An important difference between Landau and Lozano's research and the discrete augmented error method is the complete discrete-time approach, without the background of an equivalent continuous-time scheme. As will be shown, this sometimes makes it less easy to see 'what happens' in the derivations. The ITR structure resembles some self-tuning controllers (Egardt, 1980; Johnson, 1980).

4.6.1 The primary ITR controller

To derive the ITR controller, a more general discrete notation for the process will be used:

$$A(z^{-1})y_p(k) = z^{-d}B(z^{-1})u(k)$$

where:

$$A(z^{-1}) = 1 + a_1 z^{-1} + \ldots + a_n z^{-n}$$
$$B(z^{-1}) = b_0 + b_1 z^{-1} + \ldots + b_m z^{-m}$$

The process numerator and denominator orders (m and n respectively), and the time delay d, are assumed to be known.

The control goals are twofold. First, in tracking, the transfer $y_p(k)/r(k)$ of the controlled process should be equal to the reference model:

$$A_m(z^{-1}) y_m(k) = z^{-d} B_m(z^{-1}) r(k)$$

Second, in regulation ($r(k)$ = constant), an initial disturbance of the output error $e_1(k) = y_p(k) - y_m(k)$ should be eliminated with dynamics specified by:

$$C(z^{-1}) e(k+d) = 0, \quad \forall\, k \geq 0$$

Now the name 'independent tracking and regulation' is explained: the user can specify the desired tracking properties with $A_m(z^{-1})$ and $B_m(z^{-1})$, and the regulation objectives with $C(z^{-1})$. If the process parameters a_i, b_i and d are all known, it can be shown (Landau and Lozano, 1981) that these control objectives can be fulfilled by applying the controller structure of figure 4.13. The reference model *without the time delay d*, and the regulation filter $C(z^{-1})$ are placed in cascade with the process. The controller polynomials $Q(z^{-1})$ and $R(z^{-1})$ are of appropriate orders and their coefficients are the controller parameters. Note

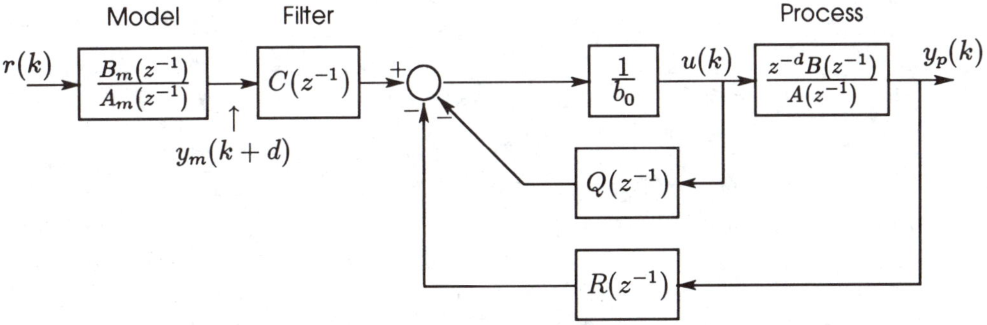

Figure 4.13 Primary controller of the ITR algorithm

that, ideally, $y_p(k) = y_m(k)$ and therefore $C(z^{-1})$ will appear as $1/C(z^{-1})$ in the loop transfer function, thereby determining the regulation behaviour. The correct controller parameters in R and Q thus depend on C. Note that C can only slow down the regulation, and hence C should only be included if the result with $C = 1$ yields a 'wild' control signal u. The control signal $u(k)$ is calculated as:

$$\begin{aligned} u(k) &= \frac{1}{b_0}\left[C(z^{-1})y_m(k+d) - Q(z^{-1})u(k) - R(z^{-1})y_p(k)\right] \\ Q(z^{-1}) &= q_1 z^{-1} + \ldots + q_{n_Q} z^{-n_Q} \\ R(z^{-1}) &= r_0 + r_1 z^{-1} + \ldots + r_{n_R} z^{-n_R} \end{aligned}$$

or alternatively:

$$u(k) = \frac{1}{b_0}\left[C(z^{-1})y_m(k+d) - \boldsymbol{\theta}_0^T \boldsymbol{\omega}_0(k)\right] \tag{4.12}$$

with $\boldsymbol{\theta}_0$ and $\boldsymbol{\omega}_0$ the parameter vector and signal vector respectively:

$$\begin{aligned} \boldsymbol{\theta}_0^T &= \left[q_1,\ \ldots,\ q_{n_Q},\ r_0,\ \ldots,\ r_{n_R}\right] \\ \boldsymbol{\omega}_0^T(k) &= \left[u(k-1),\ \ldots,\ u(k-n_Q),\ y_p(k),\ \ldots,\ y_p(k-n_R)\right] \end{aligned}$$

Adding b_0 to $\boldsymbol{\theta}_0$ and $u(k)$ to $\boldsymbol{\omega}_0(k)$ results in the vectors $\boldsymbol{\theta}$ and $\boldsymbol{\omega}(k)$:

$$\begin{aligned} \boldsymbol{\theta}^T &= \left[b_0,\ \boldsymbol{\theta}_0^T\right] \\ \boldsymbol{\omega}^T(k) &= \left[u(k),\ \boldsymbol{\omega}_0^T(k)\right] \end{aligned}$$

From equation (4.12) it follows that:

$$\begin{aligned} b_0 u(k) &= C(z^{-1})y_m(k+d) - \boldsymbol{\theta}_0^T \boldsymbol{\omega}_0(k) \\ \Longrightarrow b_0 u(k) + \boldsymbol{\theta}_0^T \boldsymbol{\omega}_0(k) &= C(z^{-1})y_m(k+d) \\ \Longrightarrow \boldsymbol{\theta}^T \boldsymbol{\omega}(k) &= C(z^{-1})y_m(k+d) \end{aligned} \tag{4.13}$$

If the process parameters a_i, b_i and d are known, the controller parameter vector $\boldsymbol{\theta}$ can be determined analytically, such that both the tracking and the regulation conditions are satisfied. The controller structure in figure 4.13 differs from the augmented error scheme mainly in the position of the reference model, the presence of the prefilter $C(z^{-1})$, and the polynomials $R(z^{-1})$ and $Q(z^{-1})$, taking the place of the auxiliary signal generators in the augmented error method. Note that, now, the reference model relative degree is not necessarily equal to that of the process.

4.6.2 Adaptive ITR

The next step in the ITR design is extension by an adaptive mechanism, which consists of a direct least-squares parameter estimator (equations (4.10) and (4.11)) with a signal vector $\boldsymbol{\omega}(k)$ and the parameter vector $\boldsymbol{\theta}(k)$. This scheme updates the adaptation gain matrix $\boldsymbol{\Gamma}(k)$ and parameter vector $\boldsymbol{\theta}(k)$ recursively. For the least-squares adaptation to be asymptotically stable the following error equation is required:

$$\epsilon(k) = H(z^{-1})\,[\boldsymbol{\theta}^* - \boldsymbol{\theta}(k)]^T\,\boldsymbol{\omega}(k-d)$$

The linear part of the error equation, $H(z^{-1})$, must be 'super'-SPR, its required class $L(\lambda)$ depending on λ_2. If $H(z^{-1}) = 1$, a common situation, then:

$$\epsilon(k) = [\boldsymbol{\theta}^* - \boldsymbol{\theta}(k)]^T\,\boldsymbol{\omega}(k-d)$$

As can be seen from equation (4.13) with $\boldsymbol{\theta} = \boldsymbol{\theta}^*$ which yields $y_p = y_m$, by definition:

$$C(z^{-1})y_p(k) = \boldsymbol{\theta}^{*T}\boldsymbol{\omega}(k-d)$$

Now, from (Landau and Lozano, 1981) the identity follows:

$$C(z^{-1})y_m(k) = \boldsymbol{\theta}^T(k-d)\boldsymbol{\omega}(k-d)$$

This equation is similar to equation (4.13) for the fixed-parameter case. $\epsilon(k)$ can be found to be:

$$\epsilon(k) = C(z^{-1})\,[y_p(k) - y_m(k)] + \bar{\epsilon}(k) \tag{4.14}$$

with:

$$\bar{\epsilon}(k) = [\boldsymbol{\theta}(k-d) - \boldsymbol{\theta}(k)]^T\,\boldsymbol{\omega}(k-d) \tag{4.15}$$

For the adaptation, $\epsilon(k)$ is calculated as:

$$\epsilon(k) = C(z^{-1})y_p(k) - \boldsymbol{\theta}^T(k)\boldsymbol{\omega}(k-d)$$

which follows from equations (4.14) and (4.15). The term $\bar{\epsilon}$ vanishes as adaptation proceeds (Landau and Lozano, 1981), and therefore $y_p(k)$ tends to $y_m(k)$ if $\epsilon(k)$ tends to zero. To calculate $\epsilon(k)$, $\boldsymbol{\theta}(k)$ is required, but to calculate $\boldsymbol{\theta}(k)$, $\epsilon(k)$ is needed: this is perhaps the best illustration for the need of an *a priori/a posteriori* correction. The error $\epsilon(k)$ can be calculated by using known signals up to $t = k - 1$:

$$\begin{aligned}
\epsilon(k) &= C(z^{-1})y_p(k) - \boldsymbol{\theta}^T(k)\boldsymbol{\omega}(k-d) \\
&= C(z^{-1})y_p(k) - \boldsymbol{\theta}^T(k-1)\boldsymbol{\omega}(k-d) + [\boldsymbol{\theta}(k-1) - \boldsymbol{\theta}(k)]^T \boldsymbol{\omega}(k-d) \\
&= C(z^{-1})y_p(k) - \boldsymbol{\theta}^T(k-1)\boldsymbol{\omega}(k-d) - [\boldsymbol{\Gamma}(k)\boldsymbol{\omega}(k-d)\epsilon(k)]^T \boldsymbol{\omega}(k-d)
\end{aligned}$$

and so:

$$\epsilon(k) = \frac{C(z^{-1})y_p(k) - \boldsymbol{\theta}^T(k-1)\boldsymbol{\omega}(k-d)}{1 + \boldsymbol{\omega}^T(k-d)\boldsymbol{\Gamma}(k)\boldsymbol{\omega}(k-d)} \tag{4.16}$$

The required error signal $\epsilon(k)$ can thus be determined, so that the equivalent error scheme consists of:

1. A linear part with transfer $H(z^{-1})$, allowing the nonlinear part not to satisfy Popov's inequality by a certain amount (see section 4.5). Note that $\epsilon(k)$ is known, and so the linear error transfer $H(z^{-1})$ equals 1 by default. Another $H(z^{-1})$ must be implemented purposely.

2. A nonlinear part containing the least-squares adaptation scheme, in which the value of λ_2 imposes a requirement for a certain class of superstrictly positive realness $L(\lambda)$ on $H(z^{-1})$. λ_2 determines how much the parameter update law differs from the original one. If $\lambda_2 = 0$, the update is equal to the original gradient adaptation; if $\lambda_2 = 1$ the 'standard' least-squares update law appears.

This combination results in an asymptotically stable adaptive system. Note that in calculating $\epsilon(k)$, the reference model output $y_m(k)$ is not needed (4.16), although $\epsilon(k)$ is an augmented and filtered form of $e_1(k) = y_p(k) - y_m(k)$ (4.14). Because of this definition of the error signal, ITR is a parallel MRAC scheme.

4.6.3 General discrete MRAC

By extending the ITR scheme by several polynomials, other existing schemes can be obtained as special cases of the new structure. One example is making use of a filtered version of the signal vector $L^{-1}\boldsymbol{\omega}(k)$ in the adaptation. The resulting expression for $u(k)$ is then filtered by the inverse filter L:

$$u(k) = L\left\{\frac{1}{b_0}\left[CL^{-1}y_m(k+d) - \boldsymbol{\theta}_0^T(k)L^{-1}\boldsymbol{\omega}_0(k)\right]\right\}$$

The error equation becomes:

$$\epsilon(k) = HL\left[\boldsymbol{\theta}^* - \boldsymbol{\theta}(k)\right]^T L^{-1}\boldsymbol{\omega}(k-d)$$

Both the error equation and the filtering concept resemble the augmented error concept, although the primary controller structure is different. In addition to inclusion of filters $L(z^{-1})$, several other polynomials can be added, making the ITR scheme more general. By choosing specific forms for $C(z^{-1})$, $L(z^{-1})$, and these extra polynomials, several existing structures can be obtained, such as:

- The plain ITR algorithm, by choosing $L(z^{-1}) = 1$.
- Series–parallel MRAC, by choosing $C(z^{-1}) = B_m(z^{-1})$.
- Parallel MRAC with filters $L(z^{-1})$, resembling the augmented error method.
- Åström and Wittenmark's minimum-variance self-tuning controller.

Despite the generality of the method described, the process must still have all its zeros inside the unit circle to avoid instability, and simulations show a considerable ringing-like behaviour, especially during the start of the adaptation (figure 4.14). In between two sampling instants an intersampling ripple remains. This is illustrated in figure 4.14 by stretching the time scale during part of the response. The following section presents a method that avoids these problems.

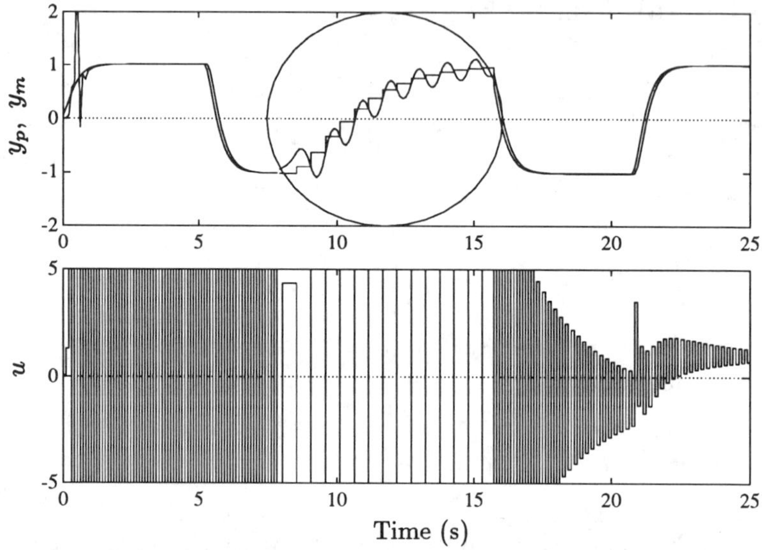

Figure 4.14 Response with standard ITR; $L = 1$, $C = 1$

4.7 Deadbeat model reference adaptive control

This section describes the design of a new MRAC scheme, making use of a 'ripple-free' or 'deadbeat' control strategy (Butler *et al.*, 1990b). The characteristic of such a controller is that the process zeros are not cancelled and thus ringing effects are avoided. In addition, nonminimum-phase processes can be controlled.

If $\boldsymbol{\theta} = \boldsymbol{\theta}^*$, the controller structure shown in figure 4.13 can be redrawn as that in figure 4.15, in which $M(z^{-1}) = z^{-d}/C(z^{-1})$. The closed-loop transfer is equal

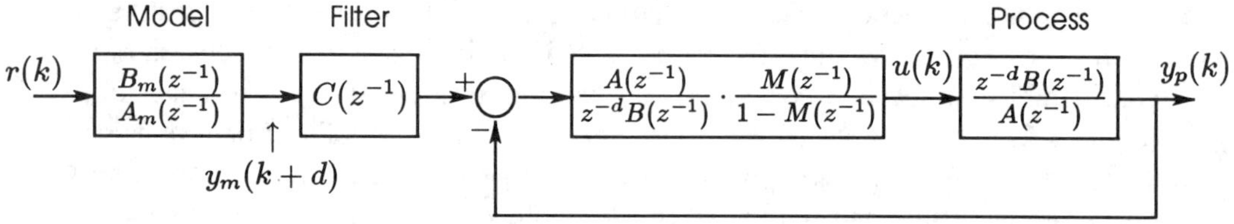

Figure 4.15 Alternative view of the primary controller of ITR

to $M(z^{-1})$ and resembles a minimum settling-time controller (MSTC; Verbruggen, 1985). A characteristic of this structure is that $u(k) = (A/B)y_m(k+d)$, and hence process zeros appear as poles in the equation for $u(k)$. After a change in

$y_m(k+d)$, $u(k+i)$ does not settle to a constant value but becomes an infinite series of control actions because of the poles in the equation for $u(k)$. On the other hand, $y_p(k) = y_m(k)$ at every instant k (but not *between* two instants). Note that for regulation, the loop transfer contains the factor $1/C(z^{-1})$, which smoothens the regulation behaviour. Because many processes have a zero on the negative real axis, MSTC usually involves the presence of ringing and intersampling ripple because it compensates all process poles and zeros. To avoid zero cancellation, M can be modified into:

$$M(z^{-1}) = \frac{z^{-d}}{C(z^{-1})} \frac{B(z^{-1})}{\sum_{i=0}^{m} b_i} = \frac{z^{-d}}{C(z^{-1})} \frac{B(z^{-1})}{b}$$

The parameter $b = \sum_{i=0}^{m} b_i$ is a compensation for the DC gain of $B(z^{-1})$. The closed-loop behaviour is known as 'ripple-free response control' (RFRC; Verbruggen, 1985) or 'deadbeat control' (Åström and Wittenmark, 1984). Because the process zeros are not cancelled, no ringing occurs. In RFRC, the control signal is $u(k) = Ay_m(k+d)$, and thus the control settles after $n+1$ steps. The process zeros remain present in $y_p(k)$:

$$y_p(k) = \frac{B(z^{-1})}{b} y_m(k) \tag{4.17}$$

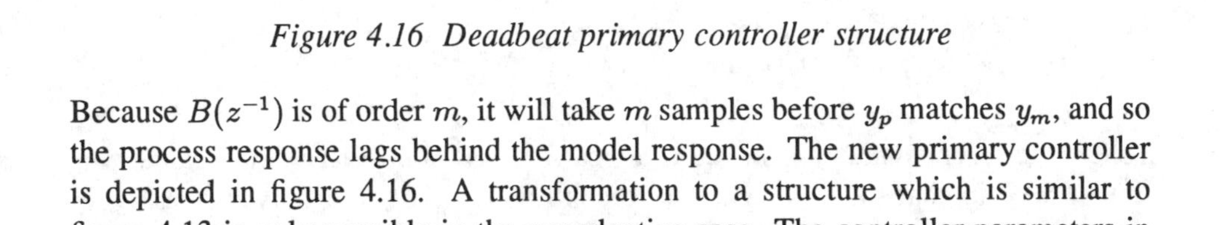

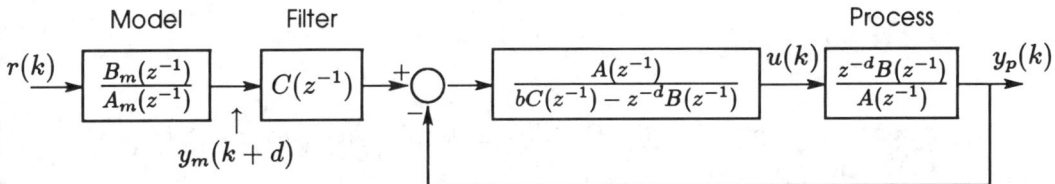

Figure 4.16 Deadbeat primary controller structure

Because $B(z^{-1})$ is of order m, it will take m samples before y_p matches y_m, and so the process response lags behind the model response. The new primary controller is depicted in figure 4.16. A transformation to a structure which is similar to figure 4.13 is only possible in the nonadaptive case. The controller parameters in figure 4.16 are calculated directly from $A(z^{-1})$ and $B(z^{-1})$, which in the adaptive case are replaced by their estimates $\widehat{A}(z^{-1})$ and $\widehat{B}(z^{-1})$. $\widehat{A}(z^{-1})$ and $\widehat{B}(z^{-1})$ are calculated with the least-squares algorithm, as in ITR. Because in this case the *process* parameters are estimated rather than the controller parameters, other definitions for $\boldsymbol{\omega}(k)$ and $\boldsymbol{\theta}(k)$ must be used. The new parameter vector $\boldsymbol{\theta}(k)$ contains estimates of the process numerator and denominator coefficients b_i and a_i, and the corresponding signal vector $\boldsymbol{\omega}(k)$ contains series of measurements of the process input and output values:

$$\begin{aligned}\boldsymbol{\theta}^T(k) &= \left[\hat{b}_0,\ \ldots,\ \hat{b}_m,\ -\hat{a}_1,\ \ldots,\hat{a}_n\right] \\ \boldsymbol{\omega}^T(k-d) &= \left[u(k-d),\ \ldots,\ u(k-d-m),\ y_p(k-1),\ \ldots,\ y_p(k-n)\right]\end{aligned}$$

Note that, due to its definition, $\boldsymbol{\omega}(k)$ cannot be generated. This is not a problem because $\boldsymbol{\omega}(k)$ is needed neither in the derivation of the adaptive law, nor in the actual control algorithm. To show that the adaptation is asymptotically stable, the following type of error equation is required:

$$\begin{aligned}\epsilon(k) &= \left[\boldsymbol{\theta}^* - \boldsymbol{\theta}(k)\right]^T \boldsymbol{\omega}(k-d) \\ &= \boldsymbol{\theta}^{*T}\boldsymbol{\omega}(k-d) - \boldsymbol{\theta}^T(k)\boldsymbol{\omega}(k-d) \qquad (4.18)\end{aligned}$$

The first term equals:

$$\boldsymbol{\theta}^{*T}\boldsymbol{\omega}(k-d) = z^{-d}Bu(k) - Ay_p(k) + y_p(k) = y_p(k) \qquad (4.19)$$

The required error signal for a stable adaptive law then becomes:

$$\epsilon(k) = y_p(k) - \boldsymbol{\theta}^T(k)\boldsymbol{\omega}(k-d)$$

Substituting the adaptive law (4.11) yields the *a priori / a posteriori* correction as in the ITR case:

$$\epsilon(k) = \frac{y_p(k) - \boldsymbol{\theta}^T(k-1)\boldsymbol{\omega}(k-d)}{1 + \boldsymbol{\omega}^T(k-d)\boldsymbol{\Gamma}(k)\boldsymbol{\omega}(k-d)}$$

Hence, the error signal $\epsilon(k)$ in the deadbeat MRAC controller is calculated in a similar way to that in the original ITR controller (4.16); the only difference is the exclusion of $C(z^{-1})$ in the calculation of $\epsilon(k)$, which is due to the different definitions of $\boldsymbol{\omega}(k)$ and $\boldsymbol{\theta}(k)$. $C(z^{-1})$ does not influence the estimated parameters in the deadbeat MRAC scheme, but is explicitely present in the primary controller.

An interesting topic is the actual meaning of the error signal $\epsilon(k)$. In the standard ITR case, $\epsilon(k)$ was: $\epsilon(k) = C(z^{-1})\left[y_p(k) - y_m(k)\right] + \bar{\epsilon}(k)$, in which

$\bar{\epsilon}(k)$ vanishes as adaptation converges. This implies that after the adaptation, $y_p(k) = y_m(k)$. In the RFRC controller, $\epsilon(k)$ can be formulated as follows.

The control signal is calculated as:

$$\begin{aligned} u(k) &= \frac{\widehat{A}}{\hat{b}C - z^{-d}\widehat{B}}\left[Cy_m(k+d) - y_p(k)\right] \qquad (4.20) \\ \Longrightarrow \hat{b}Cu(k) &= z^{-d}\widehat{B}u(k) - \widehat{A}y_p(k) + \widehat{A}Cy_m(k+d) \\ &= \boldsymbol{\theta}^T(k)\boldsymbol{\omega}(k-d) - y_p(k) + \widehat{A}Cy_m(k+d) \end{aligned}$$

And so the second term in equation (4.18) is equal to:

$$\boldsymbol{\theta}^T(k)\boldsymbol{\omega}(k-d) = y_p(k) + \hat{b}Cu(k) - \widehat{A}Cy_m(k+d) \qquad (4.21)$$

Combining equations (4.18), (4.19) and (4.21) yields:

$$\begin{aligned} \epsilon(k) &= y_p(k) - \left\{y_p(k) + \hat{b}Cu(k) - \widehat{A}Cy_m(k+d)\right\} \\ &= \widehat{A}Cy_m(k+d) - \hat{b}Cu(k) \\ &= \frac{CA}{B}\left[\frac{\widehat{A}}{A}By_m(k+d) - \hat{b}\frac{B}{A}u(k)\right] \qquad (4.22) \end{aligned}$$

Substituting equation (4.20) in equation (4.22) gives:

$$\epsilon(k) = \frac{C\widehat{A}}{\hat{b}C - z^{-d}\widehat{B}}\left[\hat{b}y_p(k) - \widehat{B}y_m(k)\right]$$

Hence, $\epsilon(k)$ is a filtered version of a modified process-model error. As adaptation converges, $\widehat{A} \rightarrow A$ and $\hat{b} \rightarrow b$ if $\boldsymbol{\omega}(k)$ is persistently exciting, and so $\epsilon(k)$ is the filtered difference between $by_p(k)$ and $By_m(k)$. If $\epsilon(k) \rightarrow 0$, $y_p(k) \rightarrow (B/b)y_m(k)$, which indeed implies deadbeat behaviour. Hence, where $\epsilon(k) \rightarrow 0$ involved $y_p(k) \rightarrow y_m(k)$ in ITR, in the new deadbeat controller $y_p(k) \rightarrow (B/b)y_m(k)$.

To conclude, a change from a minimum settling-time controller in the original ITR algorithm to a deadbeat controller structure involves a redefinition of $\boldsymbol{\theta}(k)$ and $\boldsymbol{\omega}(k)$, such that the estimated process parameters are contained in $\boldsymbol{\theta}(k)$ and the process input and output signals are present in $\boldsymbol{\omega}(k)$. From the required error equation needed for an asymptotically stable adaptive scheme, an error signal

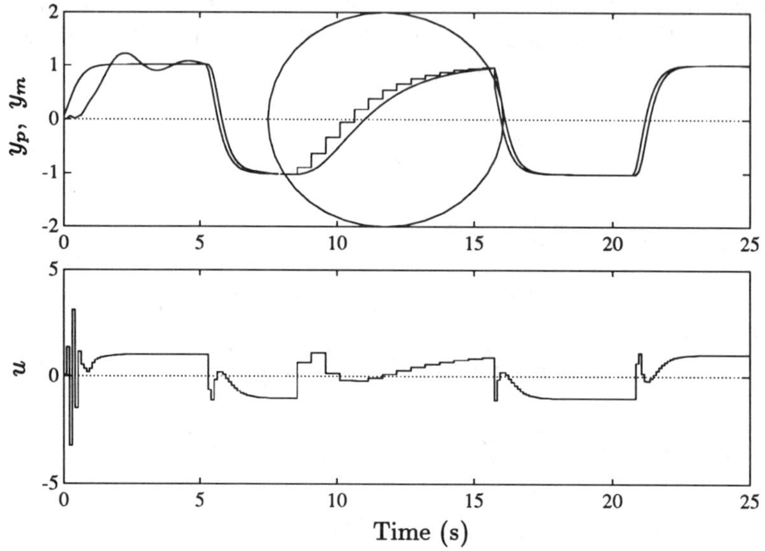

Figure 4.17 Response obtained by using the deadbeat model reference adaptive controller

evolves which imposes deadbeat behaviour in the converged state.

Simulations show that the ringing effect has disappeared completely and the control signal is much smoother. This is a well-known characteristic of deadbeat controllers (Åström and Wittenmark, 1984). In figure 4.17, a magnification similar to that in figure 4.14 has been included.

An additional advantage of the new controller structure is its ability to control nonminimum-phase processes. This has become possible because the adaptive scheme does not attempt to compensate existing process zeros. Using the same reference model, application of the control scheme to the nonminimum-phase process:

$$W_p(s) = \frac{-4s+8}{s^2+2s+8}$$

gives the response shown in figure 4.18. The standard ITR scheme in this case becomes unstable immediately. Of course, the nonminimum-phase response characteristics remain because the process zeros are not cancelled.

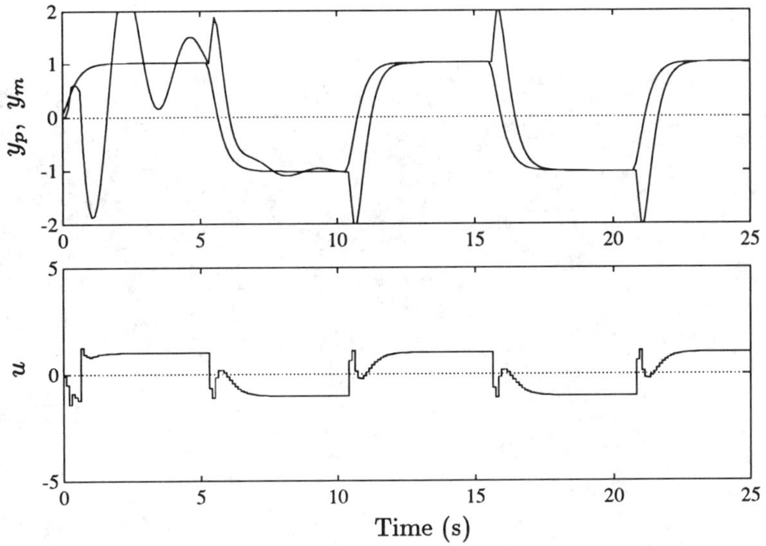

Figure 4.18 Deadbeat MRAC applied to a nonminimum-phase process

4.8 Practical results with a water-level system

The new MRAC scheme has been tested on a fluid-level process, which is shown in figure 4.19. A schematic drawing is depicted in figure 4.20, and the system can be described as follows. Out of a container, in the lower part of the setup, water is pumped, via pipes and valves, into the water column. Three valves influence the water flows. Valve S3 is the control valve controlling the water input to the column. This valve is connected, via an electropneumatic transducer, to the control computer: a VAXstation II minicomputer running MUSIC (MUlti-purpose SImulation and Control package). Valve S2 is necessary for the pump to operate properly when S3 is closed. Valve S1 determines the outflow of water from the column, and the water level in the column is measured by a pressure sensor at the bottom. A model of this process can be derived as follows.

The water flow into the column is determined by valve S3, which is controlled by the control voltage U. The relation between U and the inflow Φ_{in} can be approximated by:

$$\Phi_{\text{in}} = \frac{Ke^{-sT_d}}{s\tau + 1}U$$

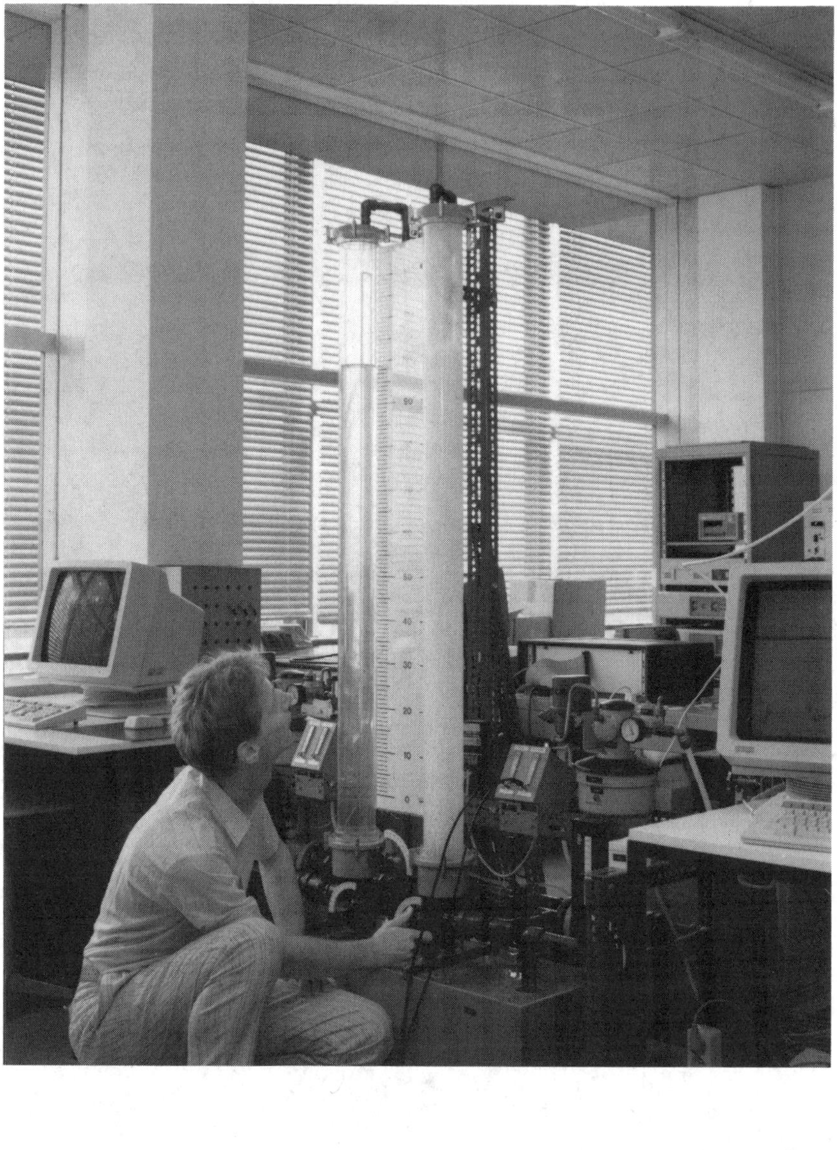

Figure 4.19 The water-level process. The setup consists of two separate processes, which can be coupled by opening a valve at the bottom. Only one of the two processes is used in the experiments

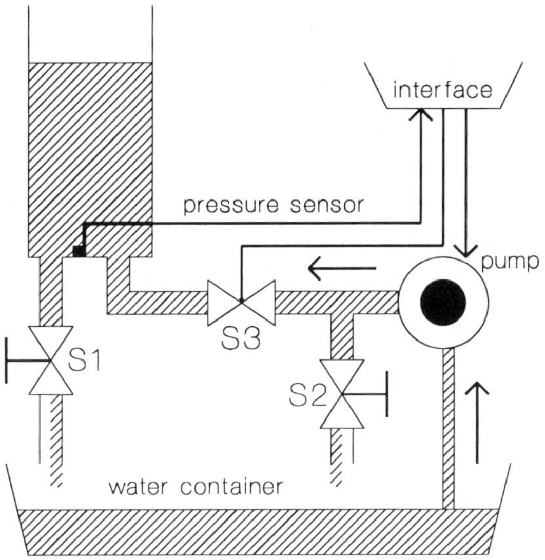

Figure 4.20 Schematic drawing of the water-level process

with approximate values of $T_d = 0.65$ s. and $\tau = 0.5$ s. The value of K is unknown but is of order 10^{-4} m^3/Vs. The outflow Φ_{out} depends on the water level and the cross section of the tubes, and can be calculated using Bernoulli's law (Alonso and Finn, 1980). Bernoulli's law expresses the conservation of energy in a fluid:

$$\frac{1}{2}\rho v_1^2 + \rho g h_1 + p_1 = \frac{1}{2}\rho v_2^2 + \rho g h_2 + p_2$$

In this equation, v_1 and v_2 are the fluid's velocity in two locations in a tube; ρ is the fluid density; h_1 and h_2 are the heights of the two locations; and p_1 and p_2 are the fluid pressures at the two locations. The total energy per volume unit, consisting of kinetic energy, potential energy and 'pressure energy', is preserved throughout the fluid. This law can be applied in the case of the water column by considering the relevant quantities at $h = h_1$ and at $h = 0$ (see figure 4.21). Note, however, that two requirements for the application of Bernoulli's law are not met, namely the absence of friction between the fluid and the pipes, and the stationarity condition (stating that all velocities must be constant). Hence, Bernoulli's law gives only an approximation of the actual process behaviour.

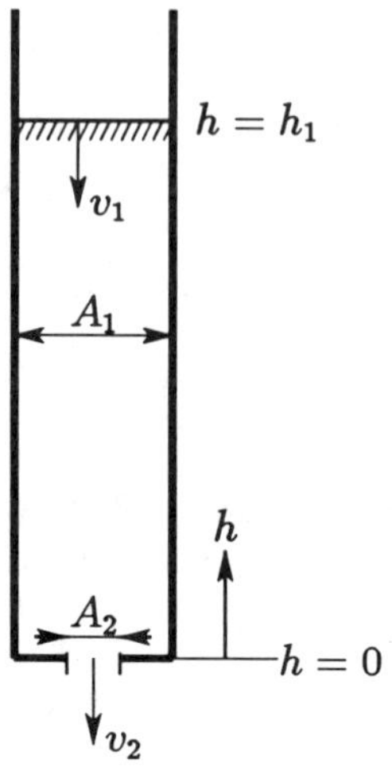

Figure 4.21 Quantities used for the derivation of the outflow

At $h = h_1$, the fluid velocity equals v_1 and the pressure p_1 equals the atmospheric pressure p_{atm}. At $h = 0$, the velocity v_2 is the velocity of the outflowing water after it has passed the opening at the bottom of the tube, and the pressure p just after the opening also equals the atmospheric pressure p_{atm}. Therefore, according to Bernoulli's law:

$$\frac{1}{2}\rho v_1^2 + \rho g h_1 + p_{\mathrm{atm}} = \frac{1}{2}\rho v_2^2 + p_{\mathrm{atm}} \tag{4.23}$$

At the same time, the ratio between the water velocity in the tube and that just after passing the opening is given by:

$$v_2 = \frac{A_1}{A_2} v_1 \tag{4.24}$$

with A_1 the section surface of the tube, and A_2 the section surface of the opening at the bottom. Substituting equation (4.24) into equation (4.23) yields:

$$\begin{aligned} \frac{1}{2}\rho\left(v_1^2 - \frac{A_1^2}{A_2^2}v_1^2\right) &= -\rho g h_1 \\ \Longrightarrow \frac{1}{2}v_1^2 &= \frac{A_2^2}{A_1^2 - A_2^2} g h_1 \end{aligned}$$

Hence, the fluid velocity in the tube equals:

$$v_1 = \sqrt{\frac{2A_2^2}{A_1^2 - A_2^2} g h_1}$$

Measurements have yielded $A_1 = 6 \times 10^{-3}$ m^2. The value of A_2 depends on the setting of valve S1, and has a maximum value of 6×10^{-4} m^2. The outflow Φ_{out} equals $A_1 v_1$, and is thus a nonlinear function of A_1 and h_1. Assuming A_2 constant during operation, the outflow can be written as:

$$\Phi_{\text{out}} = C\sqrt{h_1}$$

in which the constant C equals:

$$C = A_1 \sqrt{\frac{2A_2^2}{A_1^2 - A_2^2} g}$$

The difference between the inflow and the outflow is integrated to obtain the actual fluid level h:

$$h = \frac{\Phi_{\text{in}} - \Phi_{\text{out}}}{A_1 s}$$

This gives the overall transfer function from control voltage U to the water level h:

$$h = \frac{\frac{K e^{-sT_d}}{s\tau + 1} U - C\sqrt{h}}{A_1 s}$$

This leads to the block diagram of figure 4.22. The water level h is measured by a pressure transducer at the bottom of the tube, which has some additional, unknown, dynamics, which are neglected. In addition, the effect of the water pressure on the pump operation is neglected. In the actual system, the maximum water level is $h_{\text{max}} = 120$ cm.

A linearization of the transfer function $\Phi_{\text{out}} = C\sqrt{h}$ around a working point h_0 yields:

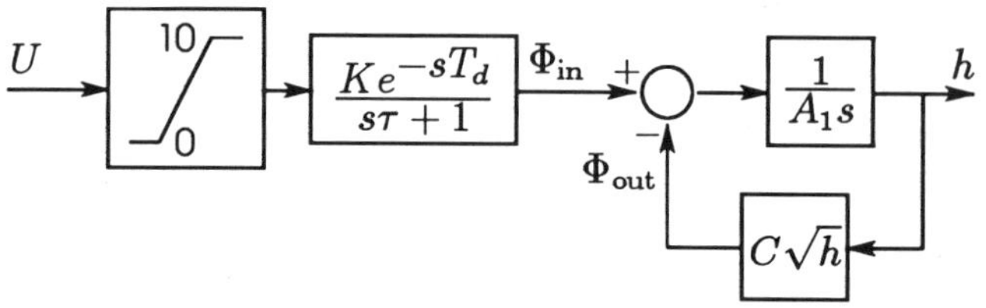

Figure 4.22 Block diagram of the water-level system

$$\Phi_{\text{out}} \approx C\sqrt{h_0} + \frac{C}{2\sqrt{h_0}}(h - h_0)$$

Hence, in the linearized system a feedback is present which is inversely proportional to the square root of the water level h. This means that the time constant depends proportionally on the square root of h.

The outflow depends heavily on the opening surface A_2, and hence the setting of the outflow valve S1 has considerable effects. The voltage U is limited to between 0 and 10 V, causing an asymmetry in the response between the rising and dropping rate of the water level. Note that a too large outflow opening A_2 may result in inability to reach a high-level set point.

The resulting model of the water-level system thus consists of a nonlinear, second-order system in which additional dynamics, for example in the pressure sensor, are neglected. In controlling the system, in the deadbeat MRAC scheme a second-order, linear reference model is used. The model is chosen such that at nominal outflow the process can just follow the model. A sampling time of 0.5 s is applied and the forgetting factor $\lambda_1 = 0.99$, while $\lambda_2 = 1$. Figure 4.23 shows the response with the deadbeat MRAC scheme. For the first 50 s the set point is chosen at 70 cm, and after this time a block-type set point is applied with a minimum of 30 cm and a maximum of 110 cm. It is observed that after initial start-up overshoot, the response approaches the reference model response. Figure 4.24 shows another response. In this case, at $t = 75$ s valve S1 is opened more, such that the outflow increases. It can be seen that the response approaches the model response after some time. At $t = 150$ s, the opening of valve S1 is made smaller, limiting the outflow. This setting prohibits perfect model-following at a negative set-point change, but the adaptation recovers quickly. Hence, the practical useability of the deadbeat MRAC scheme is demonstrated.

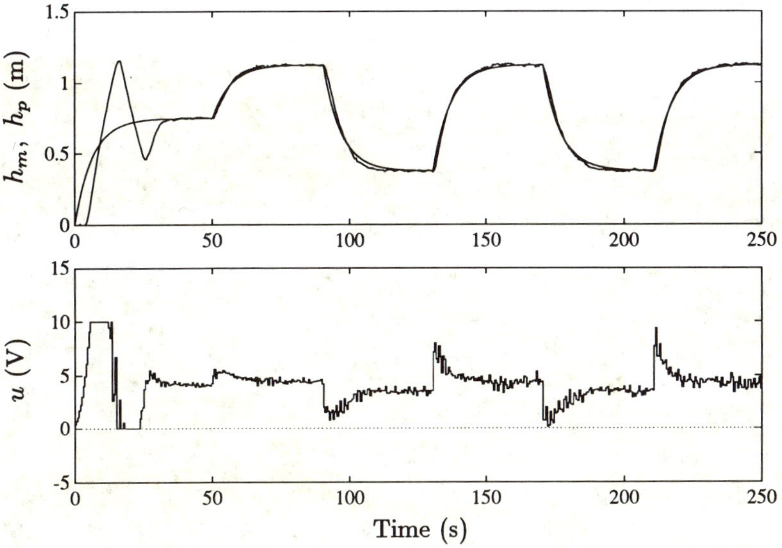

Figure 4.23 Tracking properties of deadbeat MRAC, applied to the water-level system

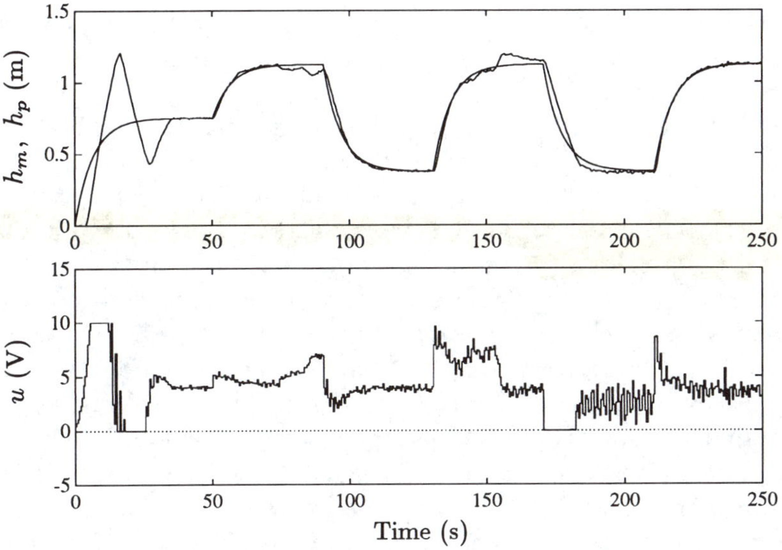

Figure 4.24 Deadbeat MRAC, applied to the water-level system: at t = 75 s *the outflow is increased, and at* t = 150 s *the outflow is decreased*

4.9 Summary

The transfer from continuous-time to discrete-time MRAC is possible due to the availability of discrete versions of the stability theories. The inherent one-sample delay in discrete MRAC causes a discrepancy between the signals available and those required to ensure a stable adaptive scheme. This can be solved by applying a correction factor which links the *a posteriori* to the *a priori* process-model error. This correction appears in all discrete MRAC algorithms.

Implementing MRAC in a discrete manner makes it possible to apply least-squares adaptation instead of gradient adaptation. Least-squares adaptation adjusts the adaptation gain matrix $\boldsymbol{\Gamma}(k)$ online, which alleviates the user's job of choosing $\boldsymbol{\Gamma}$ himself. However, least-squares adaptation introduces other tuning factors such as the forgetting factor λ_1, the factor λ_2, a starting value $\boldsymbol{\Gamma}(0)$, and parameters used for the necessary limitation of $\boldsymbol{\Gamma}(k)$. Consequently, from a user's point of view, least-squares adaptation makes life neither easier nor more difficult.

The introduction of least-squares adaptation requires a modified hyperstability theory because the algorithm violates the Popov demand on the nonlinear part of the equivalent error feedback scheme. Modified hyperstability allows this violation, but imposes heavier ('super-SPR') demands on the linear part.

The first discrete-time MRAC schemes, such as the augmented error method, were based on existing continuous-time equivalents. In the discrete augmented error method, reduction of the complexity of the auxiliary signal generators is possible, which makes the tuning of the ASGs easier. An extension by least-squares adaptation is straightforward.

The 'independent tracking and regulation' algorithm, presented by Landau and Lozano, is an example of a complete discrete-time design. In these schemes, a least-squares algorithm is usually applied instead of a gradient-like scheme. The least-squares algorithm can be shown to yield a stable adaptive system if the error equation is of a specific form. A major part of the analysis of the adaptive system therefore consists of showing that if the error signal used in the adaptation (and imposed by the least-squares method) goes to zero, then the desired closed-loop behaviour occurs. Analyzing the primary controller structure of the ITR scheme, it can be observed that the closed loop approaches a minimum-settling-time behaviour after adaptation. This may cause severe 'ringing' problems. Altering the primary controller structure to a deadbeat controller gives rise to a new definition of the parameter vector $\boldsymbol{\theta}$ and the signal vector $\boldsymbol{\omega}$. It can be shown that the error signal used in the adaptation is a filtered form of the desired error signal. When adaptation converges, the closed loop actually approaches a deadbeat transfer function. The new deadbeat scheme appears to yield satisfactory results and allows controlling nonminimum-phase processes.

Problems

4.1 In section 4.2, it was mentioned that the *a priori*/*a posteriori* correction resembles the normalization which is sometimes present in continuous-time MRAC schemes. In terms of maintaining stability, can you think of similarities in the *function* of these error modifications?

4.2 In the example given in section 4.4.2 using the discrete augmented error method, the bandwidth of the filters L^{-1} is doubled which is claimed to improve the convergence properties if gradient adaptation is used. Later, in applying least-squares adaptation, the original filters are used. Now, the linear part of the error transfer function with the 'fast' filters:

$$W_m L = 0.358 \frac{z^2 - 0.899z + 0.202}{z^2 - 1.341z + 0.449}$$

involves a frequency plot such as that shown in figure 4.10 (a), with a real part of 1 for $\omega = 0$ and a real part of 0.25 for $\omega = \omega_{\max} = 10\pi$ rad/s. Can you explain why the combination of the 'fast' filters and least-squares adaptation may result in unstable behaviour?

4.3 Would you consider the deadbeat MRAC scheme, presented in section 4.7, a *direct* or an *indirect* scheme? In what respect is this approach 'special'?

5

Structured Unmodelled Dynamics in MRAC

This chapter is dedicated to the problem of unmodelled dynamics in model reference adaptive control systems. The effects of these unmodelled dynamics are often referred to as 'state-dependent disturbances', in contrast to external disturbances which were considered in chapter 3. The main difference between these two is that state-dependent disturbances usually cannot be assumed to be bounded, whereas external disturbances can. Usually, the unmodelled dynamics lie in a frequency range outside the nominal process bandwidth because, especially in the high-frequency range, structural knowledge of the dynamics is missing. If the unmodelled dynamics have only a relatively small effect on the process output, simple adaptive law modifications are sufficient to deal with them. However, if these dynamics have a more fundamental influence on the process output, or lie in the same frequency range as the process dynamics, special precaution must be taken.

5.1 Instability caused by unmodelled dynamics

The main interest in the robustness of model reference adaptive control systems with respect to unmodelled process dynamics was instigated by Rohrs *et al.* (1985). An excellent analysis can also be found in (Sastry and Bodson, 1989). Rohrs' example consisted of a simple first-order process with unmodelled dynamics, controlled by an MRAC system which was designed only for the nominal process. It was shown that even if the unmodelled dynamics play a role only in a frequency area outside the nominal process bandwidth, and the poles of the unmodelled part are properly damped, instability may occur. This example aroused a large interest in the mechanisms playing a role in this instability. It was found that two mechanisms play a vital role (Åström, 1983; Sastry and Bodson, 1989).

First, if the reference signal r does not have sufficient richness, ambiguity in the controller parameters may remain. If there is some measurement noise on the process output, parameter drift may occur, as described in chapter 3. If there are no unmodelled dynamics this drift will usually not impose many problems other than the fact that the adaptation has to start again after a set-point change. If there are unmodelled dynamics, however, the parameter drift may easily induce instability in the system, as illustrated in example 5.1

Example 5.1 Slow drift instability

Assume a first-order process $1/(s+1)$ being controlled by two parameters k_0 and d_0, as shown in figure 5.1. Further, assume that the reference model has a DC

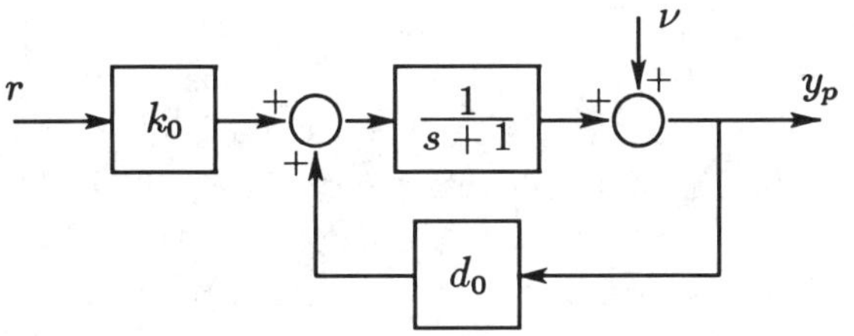

Figure 5.1 Example 5.1, *first-order process with measurement noise*

gain of 1, and that $r = 1$. The DC gain of the closed-loop system is equal to $k_0/(1 - d_0)$. Because the reference signal is constant (and hence its degree of PE is too low to identify both parameters), the measurement noise ν will cause the parameters to drift over a line in parameter space specified by $k_0 = 1 - d_0$. This is illustrated in figure 5.2 (a). Now, suppose that the process is not really of first order, but contains two extra poles somewhere near the negative real axis,

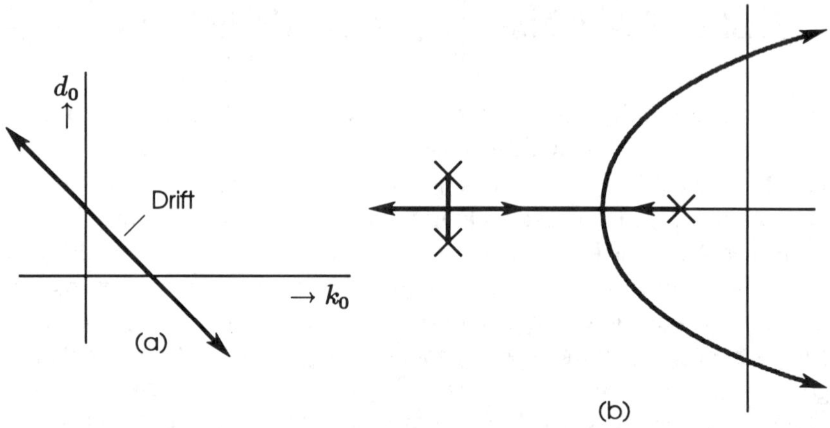

Figure 5.2 Example 5.1, *parameter drift* (a) *and root locus with process including unmodelled dynamics* (b)

as shown in figure 5.2 (b). This diagram shows how the root locus evolves as a function of the loop gain. Obviously, parameter drift of d_0 may increase the loop gain such that the system becomes unstable due to the extra poles, even though these are properly damped.

□

This mechanism is called *slow drift instability*, and is caused by a combination of low richness of the reference signal and some measurement noise. A detailed analysis of this instability mechanism using averaging techniques can be found in (Sastry and Bodson, 1989).

Second, a reference signal which is exciting in the frequency area in which the unmodelled dynamics play a role causes the adaptation to attempt model matching in this high-frequency area. However, the primary controller is not equipped to achieve this model matching, and therefore instability may result. This mechanism is of particular interest later in this chapter because, there, the frequency range in which the unmodelled dynamics are active is in the same range as the bandwidth of the nominal process.

5.2 Robustness in the presence of unmodelled dynamics

The robustness problem in the case of state-dependent disturbances is fundamentally different from that discussed in chapter 3, where external disturbances were considered. The main difference lies in the fact that a state-dependent disturbance cannot be assumed to be bounded, and, as a consequence, also has a different effect on the error equation.

The unmodelled dynamics in the process will be described as a transfer function $\widetilde{W}_p$ which is added to the original process transfer W_p:

$$\overline{W}_p = W_p + \widetilde{W}_p$$

Here, $\overline{W}_p$ is the complete process transfer function, consisting of the nominal part W_p and an unmodelled part $\widetilde{W}_p$. This process is controlled by an MRAC system which was designed for W_p only, and so a parameter vector $\boldsymbol{\theta}^*$ and a corresponding signal vector $\boldsymbol{\omega}^*$ make the output of W_p equal to the reference model output:

$$W_p \left(\boldsymbol{\theta}^{*T} \boldsymbol{\omega}^*\right) = W_m r \tag{5.1}$$

It is assumed that the true parameter vector $\boldsymbol{\theta}^*$ in the undisturbed case still stabilizes the system if there are unmodelled dynamics. Note that the signal vector is then not equal to $\boldsymbol{\omega}^*$, but will be denoted by $\boldsymbol{\omega}$. Of course, the process output in this case does not equal the reference model output $y_m = W_m r$. A fictitious reference model $\overline{W}_m$ is defined as the closed-loop process response when the primary controller is equipped with the originally exact parameter vector $\boldsymbol{\theta}^*$:

$$\overline{W}_m r = \overline{W}_p \left(\boldsymbol{\theta}^{*T} \boldsymbol{\omega}\right) \tag{5.2}$$

Because the unmodelled dynamics in the process are not known, the fictitious reference model transfer $\overline{W}_m$ is also unknown. The main reason for defining $\overline{W}_m r$ as in equation (5.2) is that the process output can be written as (Narendra and Annaswamy, 1986b):

$$y_p = \overline{W}_m \left(r + \boldsymbol{\phi}^T \boldsymbol{\omega}\right) \tag{5.3}$$

If $\phi = 0$ and so $\boldsymbol{\theta} = \boldsymbol{\theta}^*$, equation (5.2) reappears. To analyze the effect of the unmodelled dynamics on the error equation, two situations can be distinguished.

If W_m is SPR, the output error $e_1 = y_p - y_m$ can be used directly in the adaptation in the nominal case. The error equation becomes:

$$\begin{aligned} e_1 &= y_p - y_m \\ &= \overline{W_m}\left(r + \boldsymbol{\phi}^T\boldsymbol{\omega}\right) - W_m r \\ &= \overline{W_m}\left(\boldsymbol{\phi}^T\boldsymbol{\omega}\right) + \widetilde{W_m} r \\ &= \overline{W_m}\left(\boldsymbol{\phi}^T\boldsymbol{\omega}\right) + \nu \end{aligned} \tag{5.4}$$

In equation (5.4), a disturbance $\nu = \widetilde{W_m} r$ on the output error appears, with $\widetilde{W_m} = \overline{W_m} - W_m$. This disturbance can be written in an alternative form:

$$\begin{aligned} \overline{W_m} r &= \overline{W_p}\left(\boldsymbol{\theta}^{*T}\boldsymbol{\omega}\right) \\ W_m r + \widetilde{W_m} r &= W_p\left(\boldsymbol{\theta}^{*T}\boldsymbol{\omega}\right) + \widetilde{W_p}\left(\boldsymbol{\theta}^{*T}\boldsymbol{\omega}\right) \\ W_m r + \widetilde{W_m} r &= W_p\left[\boldsymbol{\theta}^{*T}\left(\boldsymbol{\omega} - \boldsymbol{\omega}^*\right)\right] + W_p\left(\boldsymbol{\theta}^{*T}\boldsymbol{\omega}^*\right) + \widetilde{W_p}\left(\boldsymbol{\theta}^{*T}\boldsymbol{\omega}\right) \end{aligned}$$

Equation (5.1) then yields:

$$\widetilde{W_m} r = W_p\left[\boldsymbol{\theta}^{*T}\left(\boldsymbol{\omega} - \boldsymbol{\omega}^*\right)\right] + \widetilde{W_p}\left(\boldsymbol{\theta}^{*T}\boldsymbol{\omega}\right) \tag{5.5}$$

This equation shows that the disturbance on the output error (which is equal to the effect of the unmodelled dynamics on y_p for $\boldsymbol{\theta} = \boldsymbol{\theta}^*$) consists of two parts. Suppose for a moment that $\boldsymbol{\phi} = 0$, so that e_1 is equal to the output disturbance only, and is not influenced by the adaptation error (the first term in equation (5.4)). In this case, the output error e_1 is equal to equation (5.5). The first term in equation (5.5) is the output of the nominal process, with the product of the parameter vector and the difference between the *actual* signal vector $\boldsymbol{\omega}$ and the *originally perfect* signal vector $\boldsymbol{\omega}^*$ as input. Thus, the first term consists of the output of the nominal process, with a signal which is completely due to the penetration of the unmodelled dynamics in the signal vector as input. The second term in equation (5.5) corresponds to the output of the unmodelled part of the process, with the actual process input $u = \boldsymbol{\theta}^{*T}\boldsymbol{\omega}$ as input.

Example 5.2 Calculation of fictitious reference model transfer function

Assume a nominal process:

$$y_p = W_p u = \frac{1}{s+1} u$$

Controlled by a controller $u = k_0 r + d_0 y_p$ this yields a closed-loop transfer function:

$$y_p = \frac{k_0}{s+1-d_0} r$$

With a reference model $y_m = W_m r = \frac{4}{s+4}$, θ^* can easily be calculated:

$$k_0^* = 4, \quad d_0^* = -3$$

Now, assume that the real process with unmodelled dynamics is:

$$\overline{W_p} = \frac{1}{s+1} + \frac{1}{s+5} = \frac{2s+6}{s^2+6s+5}$$

Then $\overline{W_m}$ can be calculated by evaluating the closed-loop transfer function of $\overline{W_p}$ assuming that $\theta = \theta^*$:

$$\overline{W_m} = \frac{2k_0^*(s+3)}{s^2+(6-2d_0^*)s+(5-6d_0^*)} = \frac{8s+24}{s^2+12s+23}$$

And hence:

$$\begin{aligned} \widetilde{W_m} &= \overline{W_m} - W_m \\ &= \frac{8s+24}{s^2+12s+23} - \frac{4}{s+4} \\ &= 4\frac{s^2+2s+1}{s^3+16s^2+71s+92} \end{aligned}$$

This shows how $\overline{W_m}$ and $\widetilde{W_m}$ can be calculated when the process transfer function and θ^* are available. Note that the result of this calculation depends on θ^*, which in turn depends on W_m. Note also that in this example both $\overline{W_m}$ and $\widetilde{W_m}$ are SPR, which is usually not the case.

□

If W_m is not SPR, an error-augmenting network is needed and the signal vector $\boldsymbol{\omega}$ changes to $\boldsymbol{\xi} = L^{-1}\boldsymbol{\omega}$. The augmented error becomes:

$$
\begin{aligned}
\epsilon &= e_1 - W_m L \left(L^{-1}\boldsymbol{\phi} - \boldsymbol{\phi} L^{-1}\right)^T \boldsymbol{\omega} \\
&= \overline{W_m}\left(\boldsymbol{\phi}^T\boldsymbol{\omega}\right) + \widetilde{W}_m r - W_m\left(\boldsymbol{\phi}^T\boldsymbol{\omega}\right) + W_m L\left(\boldsymbol{\phi}^T\left(L^{-1}\boldsymbol{\omega}\right)\right) \\
&= \overline{W_m} L\left(\boldsymbol{\phi}^T L^{-1}\boldsymbol{\omega}\right) + \widetilde{W}_m L\left(L^{-1}\boldsymbol{\phi} - \boldsymbol{\phi} L^{-1}\right)^T \boldsymbol{\omega} + \widetilde{W}_m r \\
&= \overline{W_m} L\left(\boldsymbol{\phi}^T\boldsymbol{\xi}\right) + \nu'
\end{aligned}
\tag{5.6}
$$

Now, a disturbance ν' arises which is different from ν, due to the error-augmenting signal which is only equipped for W_m instead of $\overline{W_m}$. Comparing the error equations (5.4) and (5.6) with the original equations in chapter 2 ((2.17) and (2.22)), two effects of the unmodelled dynamics can be observed.

First, an output disturbance ν or ν' is present. If W_m is SPR, this disturbance equals $\nu = \widetilde{W}_m r$, and because it was assumed that the original exact parameter vector $\boldsymbol{\theta}^*$ stabilizes the overall system, ν is bounded. If W_m is not SPR, the disturbance ν' includes a term due to an incorrect error-augmenting signal and boundedness is not automatically guaranteed. All modifications presented in chapter 3, such as the inclusion of a dead zone or γ-modification, can be used to minimize the effect of ν. More formally, if there is an upper bound ν_0 on the disturbance ν or ν' that satisfies:

$$
\nu_0 \le \beta\left(1 + \boldsymbol{\xi}^T\boldsymbol{\xi}\right)^{\frac{1}{2}}
$$

for a sufficiently small β, all modifications of the adaptive law (see section 3.3) guarantee stability (Narendra and Annaswamy, 1986b). Thus, if the disturbance due to unmodelled dynamics is small compared to the signal vector, stability can be guaranteed by using a standard adaptive law modification. Also, if persistently exciting (PE) properties are used, stability can be proven for $\mu_0 > \rho\nu_0$ (see also section 3.3.2). In this specific case, the degree of PE μ_0 of the signal vector and the upper bound on the disturbance ν_0 are both proportional to the amplitude of r. This implies that no conditions are put on the degree of PE of r, but only on the system properties in the form of the signal generators, filters, and W_m, which all affect the value of ρ. Because this relation is unknown, an analysis with the aid of PE properties is difficult.

Second, the linear error transfer function has changed from $W_m L$ to $\overline{W_m} L$. Because $\overline{W_m} L$ includes an unknown part due to the unmodelled process dynamics, the SPR property of the error equation is endangered. This problem can be dealt with by using the theory of averaging, which provides a less strict assumption on the linear part than the original strictly positive real requirement (see section 1.3.5, equation (1.8)), under the assumption of a small adaptation gain. The linear part of the error equation only needs to have a positive real part for the most important frequencies in the signal vector. Roughly stated, the averaging method tells us that the closer $\overline{W_m}L$ is to the positive real property (i.e. the larger the frequency range over which its real part is positive), the better the stability properties of the adaptive system.

To conclude: the presence of unmodelled dynamics in the process results in two types of disturbance in the error equation. An output disturbance needs modifications of the adaptive law or requirements on the degree of PE of the signal vector ξ. In addition, the SPR property of the linear part is endangered, which may be harmless if $\mathrm{Re}[\overline{W_m}L(j\omega)]$ is positive for the most important frequencies in ξ. Note that decreasing the adaptive gains leads to a larger difference in the adaptation dynamics and the process dynamics, in turn resulting in a wider application of the averaging principle.

5.3 Reference model decomposition

Reference model decomposition is a method of improving the robustness of an MRAC system with regard to unmodelled dynamics, where there is structural knowledge of these dynamics and where there is some indication of the relevant parameter values (Butler *et al.*, 1991b). Typically, such knowledge is present in the low-frequency range. Therefore, the problem setting is different from that usually encountered in the literature, where mainly high-frequency unmodelled dynamics are considered. The decomposition method can be regarded as a way of keeping the controller order low, despite the presence of extra dynamics. The designer can decide which part of the system is regarded as 'nominal' and which part as 'unmodelled'. If the extra dynamics have a relatively small effect on the process output, a standard modification can cope with them. If the extra dynamics lie in the same frequency range as the process bandwidth and have a considerable influence on the process output, their existence is generally known, and neglecting them in the control strategy may be done purposely. The process dynamics can then be divided into three classes: the *nominal* part for which the controller is designed; the *structured unmodelled* part which is known to exist but disregarded to keep the controller order low; and the *unmodelled* part (usually in the high-frequency range) which is disregarded altogether, or for which a 'standard' modification is implemented. The decomposition method is meant for the second class, whereas in the literature the main attention is focused on robustness with respect to the third class.

5.3.1 Introduction

The idea of reference model decomposition originates from predictive control (Richalet *et al.*, 1990), in which an internal model makes an on-line prediction of the future process output. This prediction depends on the future control actions, which are to be calculated. Based on the internal model, a series of control actions in the future is calculated by minimizing a criterion function, which usually depends on the difference between the desired process response and the internal model output in the future. Also, the magnitude of the control actions themselves or, for example, the increments in the control actions, may be included (Soeterboek, 1992). To obtain good results, it is essential that the internal model closely resembles the actual process transfer function. This can be achieved by on-line process identification, leading to adaptive predictive controllers. Another method is to choose a fixed internal model and to optimize the robustness properties with respect to process-model mismatch in simulation (Richalet *et al.*, 1990). In the latter case, it has been shown that if the process is unstable or has badly damped poles, even a small process-model mismatch can prove disastrous for the performance. Intuitively this can be easily explained: even if the difference between the position of an unstable process pole and that of the corresponding model pole is very small, the difference between both responses can become unbounded. This problem has led to the decomposition method in predictive control, which effectively corrects the internal model output for the process-model mismatch (Richalet *et al.*, 1990).

Although the underlying idea is the same when applying model decomposition in model reference adaptive control, the following important differences can be noted. In predictive control, an attempt is made to fit the internal model more closely to the actual process response in order to achieve as accurate a prediction as possible. The decomposition requires knowledge of the structure and parameters of the unstable or badly damped process part. The overall controller is equipped to control the complete process, including the unstable or badly damped part, and is therefore of high order. In MRAC, the decomposition is used to correct the reference model output for unmodelled process dynamics, which may be properly damped (even then, their presence can cause instability). By using the reference model output correction, the error equation is made less sensitive to the unmodelled dynamics. This allows a simple primary controller which is only designed for the 'nominal' process part, thus reducing the number of adjustable parameters. To apply the decomposition method, knowledge about the structure of the unmodelled dynamics is required. A theoretical analysis shows the mechanisms by which the decomposition method achieves this improvement (Butler *et al.*, 1990a and 1991b).

5.3.2 Basic decomposition idea

In applying the decomposition, a system $H(s)$ to be decomposed is first separated into a nominal part $H_n(s)$ and an unmodelled part $H_u(s)$:

$$H(s) = H_n(s)H_u(s)$$

Next, $H_u(s)$ is decomposed into three polynomials $T_1(s)$, $T_2(s)$ and $N(s)$, as shown in figure 5.3.

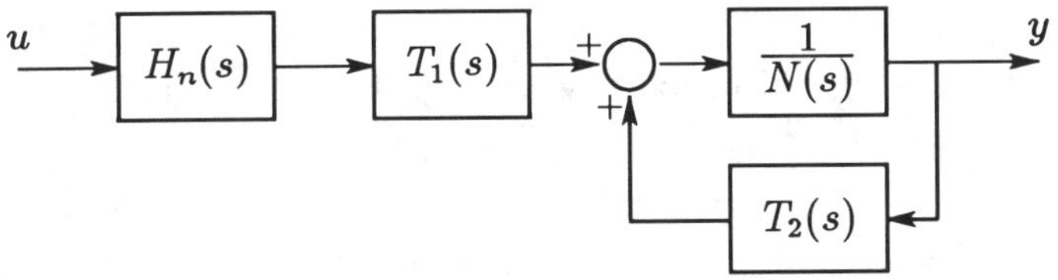

Figure 5.3 Decomposition of $H(s)$ into $T_1(s)$, $T_2(s)$ and $N(s)$

The decomposition polynomials must be chosen such that:

$$\frac{T_1}{N - T_2} = H_u$$

This leads to the observation that T_1 and $(N - T_2)$ are determined by H_u, but N may be chosen freely as long as the order of N is equal to the order of the denominator polynomial of H_u. The decomposition described is always possible.

Example 5.3 Decomposition of example 5.2

In this example we will apply decomposition to the process described in example 5.2:

$$W_p = \frac{1}{s+1}$$

$$\overline{W_p} = \frac{1}{s+1} + \frac{1}{s+5} = 2\frac{s+3}{s^2+6s+5} = \frac{1}{s+1} \cdot \frac{2s+6}{s+5}$$

The transfer $1/(s+1)$ is the nominal part of $\overline{W_p}$, making the unmodelled part:

$$H_u(s) = \frac{2s+6}{s+5} = \frac{T_1}{N - T_2}$$

Hence,

$$T_1 = 2s+6, \quad N - T_2 = s+5$$

For example, N could be chosen as $N = s+10$, which immediately gives $T_2 = 5$.

□

If H is either an internal prediction model or a reference model, the output of H is used to calculate an error signal between the process and the model. The decomposition effect is then achieved by feeding back not y_m via T_2, but y_p, as shown in figure 5.4. Here, the nominal transfer H_n consists of the original ref-

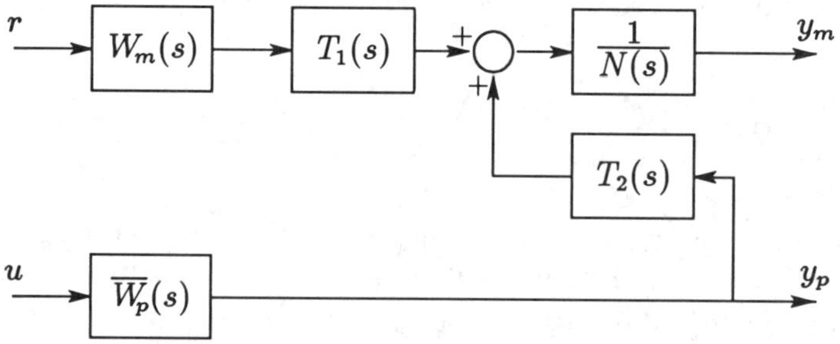

Figure 5.4 Use of the decomposition principle in MRAC

erence model W_m, and the decomposition polynomials T_1, T_2 and N perform a correction of the reference model output y_m. This correction is meant to make the reference model output have a similar 'unmodelled' part as the process output.

Intuitively, this is most clear if the polynomials T_1, T_2 and N are exactly the same as in a process decomposition (which, of course, is possible in the same way). The decomposition transfer $1/N$ is then controlled by two inputs. The first input, W_mT_1r, is similar to the corresponding input in the process decomposition, in the sense that no oscillations are present and this input is 'well behaved'. The second input, T_2y_p, represents the critical part caused by the unmodelled dynamics (which may be poorly damped oscillations). If the decomposition parameters are exact, this input is the same in the model and in the process decompositions and is thus a correction to the states in $1/N$. The method of decomposition can be regarded as a special form of model updating (van den Bosch, 1987), and has similarities with series–parallel MRAS.

The decomposition polynomials are not considered part of the reference model, but, rather, as an addition to the existing structure. The polynomials T_1, T_2 and N together form the *decomposition model.* For an exact decomposition, the polynomials T_1, T_2 and N should be the same as those that would appear in a process decomposition. However, an exact process decomposition is not available, making the choice of the decomposition parameters an interesting topic.

5.3.3 Effects of decomposition on the error equation

General effects

In the decomposition analysis also, $\boldsymbol{\theta}^*$ is defined as the parameter vector for which $W_p(\boldsymbol{\theta}^{*T}\boldsymbol{\omega}^*) = W_mr$. For this analysis, only a strictly positive real W_m is considered. Analyzing figure 5.4, the output error can be written as:

$$\begin{aligned} e_1 \;=\; y_p - y_m \;&=\; y_p - \frac{T_1}{N}W_mr - \frac{T_2}{N}y_p \\ &=\; \left(\frac{N-T_2}{N}\right)y_p - \frac{T_1}{N}W_mr \end{aligned} \tag{5.7}$$

Using equation (5.3), this yields:

$$\begin{aligned} e_1 \;&=\; \left(\frac{N-T_2}{N}\right)\overline{W_m}\left(r+\boldsymbol{\phi}^T\boldsymbol{\omega}\right) - \frac{T_1}{N}W_mr \\ &=\; \left(\frac{N-T_2}{N}\right)\overline{W_m}\left(\boldsymbol{\phi}^T\boldsymbol{\omega}\right) + \left(\frac{N-T_2}{N}\right)\overline{W_m}r - \frac{T_1}{N}W_mr \end{aligned} \tag{5.8}$$

Next, y_m^* denotes the output of the reference model corrected by the decomposition; for $\boldsymbol{\theta} = \boldsymbol{\theta}^*$:

$$\begin{aligned} y_m^* &= \frac{T_1}{N} W_m r + \frac{T_2}{N} y_p^* \\ &= \frac{T_1}{N} W_m r + \frac{T_2}{N} \overline{W}_p \left(\boldsymbol{\theta}^{*T} \boldsymbol{\omega}\right) \end{aligned} \tag{5.9}$$

Using this definition and equation (5.2), it easily follows that the error equation can be written as:

$$e_1 = \left(\frac{N - T_2}{N}\right) \overline{W}_m \left(\boldsymbol{\phi}^T \boldsymbol{\omega}\right) + \overline{W}_m r - y_m^* \tag{5.10}$$

Example 5.4 Linear part of the error equation changed by decomposition

The linear part of the error equation in the previous examples was:

$$\overline{W}_m = \frac{8s + 24}{s^2 + 12s + 23}$$

Now, making use of the decomposition proposed in example 5.3:

$$T_1 = 2s + 6, \quad T_2 = 5, \quad N = s + 10$$

the new linear part becomes:

$$\left(\frac{N - T_2}{N}\right) \overline{W}_m = \frac{s + 5}{s + 10} \cdot \frac{8s + 24}{s^2 + 12s + 23}$$

The decomposition adds one pole and one zero to the linear part of the error equation. Because the zero is situated closest to the origin in the complex plane, its effect is larger than that of the pole. Hence, the decomposition has the effect of a phase-lead compensation on the error equation.

□

Equation (5.10) shows two differences with regard to the original error equation (5.4). First, the error transfer $\overline{W}_m$ has changed to $[(N - T_2)/N]\overline{W}_m$. By choosing the decomposition parameters, the linear part of the error equation can therefore be influenced, making it possible to make this part more positive real. In other words, the frequency range in which $\mathrm{Re}[\overline{W}_m(j\omega)] > 0$ can be increased. This is advantageous because the averaging theory indicates better stability properties. Second, the output disturbance has changed from $\nu = (\overline{W}_m - W_m)r$ to $\nu = \overline{W}_m r - y_m^*$ (5.5). By the assumptions made, this latter disturbance is bounded. A further analysis of this disturbance enhances the understanding of the decomposition effects.

Effect on the output disturbance

Combining equations (5.2), (5.8) and (5.9) gives:

$$\begin{aligned}
\overline{W_m}r - y_m^* &= \overline{W_p}\left(\theta^{*T}\omega\right) - \left[\frac{T_1}{N}W_m r + \frac{T_2}{N}\overline{W_p}\left(\theta^{*T}\omega\right)\right] \\
&= \left(\frac{N-T_2}{N}\right)\overline{W_p}\left(\theta^{*T}\omega\right) - \frac{T_1}{N}W_m r \\
&= \left(\frac{N-T_2}{N}\right)\overline{W_p}\left(\theta^{*T}\omega\right) - \frac{T_1}{N}W_p\left(\theta^{*T}\omega^*\right) \\
&= \left(\frac{N-T_2}{N}\right)\overline{W_p}\left(\theta^{*T}\omega\right) - \frac{T_1}{N}W_p\left(\theta^{*T}\omega\right) + \frac{T_1}{N}W_p\left[\theta^{*T}\left(\omega-\omega^*\right)\right]
\end{aligned}$$

Using equation (5.5), this becomes:

$$\begin{aligned}
\overline{W_m}r - y_m^* &= \left(\frac{N-T_2}{N}\right)\overline{W_p}\left(\theta^{*T}\omega\right) - \frac{T_1}{N}W_p\left(\theta^{*T}\omega\right) \\
&\quad + \frac{T_1}{N}\left[\widetilde{W_m}r - \widetilde{W_p}\left(\theta^{*T}\omega\right)\right] \qquad (5.11)
\end{aligned}$$

Considering the link between the original form of describing the unmodelled dynamics by $\widetilde{W_p}$, and a process decomposition in the form of polynomials T_{p1}, T_{p2} and N_p:

$$\begin{aligned}
\overline{W_p} &= W_p + \widetilde{W_p} \\
&= W_p\left(\frac{T_{p1}}{N_p - T_{p2}}\right) \qquad (5.12)
\end{aligned}$$

it is easy to verify that:

$$W_p = \left(\frac{N_p - T_{p2}}{-N_p + T_{p1} + T_{p2}}\right)\widetilde{W_p} \qquad (5.13)$$

Using equation (5.12), equation (5.11) can be written as:

$$\begin{aligned}
\overline{W_m}r - y_m^* &= \left(\frac{N-T_2}{N}\cdot\frac{T_{p1}}{N_p - T_{p2}} - \frac{T_1}{N}\right)W_p\left(\theta^{*T}\omega\right) \\
&\quad + \frac{T_1}{N}\left[\widetilde{W_m}r - \widetilde{W_p}\left(\theta^{*T}\omega\right)\right]
\end{aligned}$$

And using equation (5.13), this finally becomes:

$$\overline{W_m r} - y_m^* = -\frac{T_{p1}}{N}\left(\frac{-N + T_1 + T_2}{-N_p + T_{p1} + T_{p2}}\right)\widetilde{W_p}\left(\theta^{*T}\omega\right) + \frac{T_1}{N}\widetilde{W_m r} \tag{5.14}$$

The meaning of equation (5.14) can be clarified as follows. Suppose that the unmodelled dynamics are known exactly, and so the decomposition polynomials T_1, T_2 and N can be chosen equal to T_{p1}, T_{p2} and N_p. Using equation (5.5), equation (5.14) then becomes:

$$\overline{W_m r} - y_m^* = \frac{T_1}{N}\left\{W_p\left[\theta^{*T}\left(\omega - \omega^*\right)\right]\right\} \tag{5.15}$$

Comparing equation (5.15) with equation (5.5), it can be seen that the second part of the disturbance ν has vanished. Therefore, a perfect decomposition removes the part from ν that corresponds with the unmodelled dynamics, which have as input the actual process input. The remaining disturbance consists of the output of a nominal process transfer function with an input that directly depends on the error in the signal vector which is caused by the unmodelled dynamics, multiplied by T_1/N.

Example 5.5 Disturbance with and without decomposition

This example studies the disturbance on the error equation for the process and decomposition given in examples 5.2 and 5.3. Recall that the disturbance without decomposition is given by $\widetilde{W_m r}$:

$$\widetilde{W_m r} = 4\,\frac{s^2 + 2s + 1}{s^3 + 16s^2 + 71s + 92}r$$

After decomposition, the disturbance is given by $\overline{W_m r} - y_m^*$. Assuming that the decomposition is exact, equation (5.14) simplifies to:

$$\overline{W_m r} - y_m^* = -\frac{T_{p1}}{N}\widetilde{W_p}\left(\theta^{*T}\omega\right) + \frac{T_1}{N}\widetilde{W_m r}$$

Using the decomposition of example 5.3:

$$T_1 = 2s + 6, \quad T_2 = 5, \quad N = s + 10,$$

this becomes:

$$\overline{W_m r} - y_m^* = \left(-\frac{2s + 6}{s + 10}\right)\widetilde{W_p}\left(\theta^{*T}\omega\right) + \left(\frac{2s + 6}{s + 10}\right)\widetilde{W_m r}$$

Now, using equations (5.12) and (5.13) it can easily be seen that:

$$\widetilde{W_p}\left(\theta^{*T}\omega\right) = \left(\frac{-N_p + T_{p1} + T_{p2}}{T_{p1}}\right)\overline{W_p}\left(\theta^{*T}\omega\right) = \left(\frac{-N_p + T_{p1} + T_{p2}}{T_{p1}}\right)\overline{W_m}r$$

Hence,

$$\begin{aligned}
\overline{W_m}r - y_m^* &= \left(-\frac{2s+6}{s+10}\cdot\frac{s+1}{2s+6}\right)\overline{W_m}r + \left(\frac{2s+6}{s+10}\right)\widetilde{W_m}r \\
&= -\frac{s+1}{s+10}\cdot\frac{8s+24}{s^2+12s+23} + 4\frac{2s+6}{s+10}\cdot\frac{s^2+2s+1}{s^3+16s^2+71s+92} \\
&= -24\frac{s^2+4s+3}{s^4+26s^3+231s^2+802s+920}
\end{aligned}$$

If N is chosen as $s+1$, T_2 vanishes and this expression simplifies to:

$$\overline{W_m}r - y_m^* = -2.4\frac{s+3}{s^3+16s^2+71s+92}$$

This illustrates that even very simple systems may have complex expressions for the disturbance. Note that the decomposition parameters were not chosen with care, and hence an analysis of the properties of these formulae is not very useful.
□

The choice of the decomposition parameters

In the choice of the decomposition parameters, there is the following dilemma. First, the linear part of the error equation is changed to $[(N - T_2)/N]\overline{W_m}$, making it possible to obtain a more positive real error equation for increased stability. Because $\overline{W_m}$ is unknown, the decomposition parameters cannot be chosen accurately to achieve this. In general, however, the poles determined by N must have a proper damping ratio and the transfer $[(N - T_2)/N]$ should have a phase lead which is as large as possible, in order to obtain the best results in this respect. This can be achieved by choosing N properly. The second decomposition effect is a decrease in the output disturbance. To remove one of the terms in the disturbance completely, the model decomposition parameters should be as close as possible to the process parameters. This still leaves the choice of N free.

Hence, the two decomposition effects result in different requirements on the decomposition parameters. To decrease the output disturbance on the error equation as much as possible, the decomposition parameters should be as close as possible to the parameters of an exact process decomposition. To obtain an error equation of which the frequency range over which its real part is positive is the largest, the choice of N is particularly important. Note that the decomposition changes the reference model output and hence the *control goal.* The analysis of the decomposition effects on the error equation, as presented above, does not disclose anything about the price to be paid in terms of performance. The freedom in choosing N must therefore not only be used to find a decomposition model that makes the output error insensitive to the unmodelled dynamics, but must also result in a desired closed-loop response. Similarly, while an exact choice of T_1 and T_2 is favourable in terms of the error equation analysis, modifications of T_1 and T_2 may be necessary to achieve a proper closed-loop behaviour. Examples of this can be found in the following sections. To conclude, although the decomposition polynomials all have a clearly distinguishable function, the choice of the decomposition parameters is not trivial and depends on the problem at hand. Some feeling for the process is needed to obtain a proper tuning, which can best be obtained with the aid of (simulation) experiments.

5.3.4 A first example

This section shows, by a simple example, the design steps to be taken in applying the decomposition method. First, the process and model transfers are given and the decomposition is applied, and next the effects on the error equation are shown. Simulations confirm the expected improvement.

Process and model transfers and decomposition

The nominal process to be considered consists of only an integrator:

$$W_p = \frac{1}{s}$$

In practice, however, there are extra first-order dynamics:

$$\overline{W_p} = \frac{1}{s(s\tau_p + 1)}$$

Because the system is primarily designed for a process of order 1, a first-order reference model is chosen:

$$W_m = \frac{1}{s+1}$$

The primary controller needs two adjustable parameters k_0 and d_0 to calculate the control signal u:

$$u = k_0 r + d_0 y_p$$

The first-order reference model yields an SPR error equation in the nominal case and therefore an error-augmenting network is not implemented. To compensate for the extra process dynamics, the decomposition is chosen to resemble a transfer $1/(s\tau_m + 1)$, yielding:

$$\frac{T_1}{N - T_2} = \frac{1}{s\tau_m + 1}$$

Choosing the polynomial $N = s\tau_n + 1$ and inspecting this equation, it can be seen immediately that:

$$\begin{aligned} T_1 &= 1 \\ N - T_2 &= s\tau_m + 1 \implies T_2 = (\tau_n - \tau_m)s \end{aligned}$$

The parameter τ_m represents the expected nominal time constant of the extra process dynamics. The design parameter τ_n can be freely chosen and thereby determines T_2 and N. Note that in the implementation, the pure differentiating terms in T_2 impose no problems because y_m is calculated as $y_m = (T_1/N)W_m r + (T_2/N)y_p$.

Analysis of the error equation

For the nominal case, $k_0^* = 1$ and $d_0^* = -1$. The actual system transfer function for these parameters is:

$$\frac{y_p}{r} = \frac{1}{\tau_p s^2 + s + 1} = \overline{W_m}$$

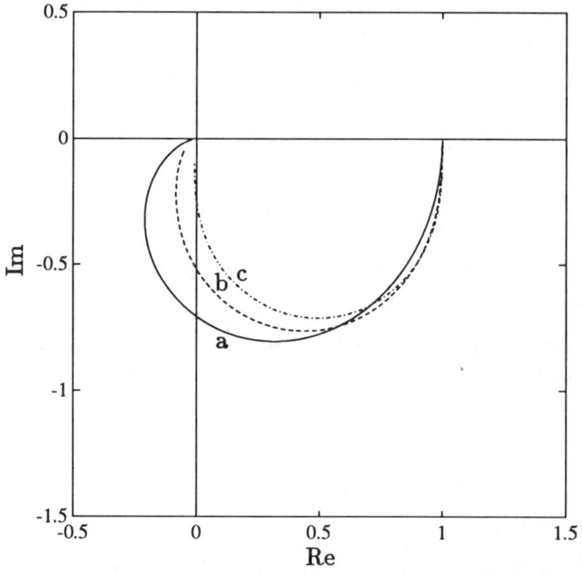

Figure 5.5 First example, Nyquist plots of $\overline{W_m}$ *and* $[(N - T_2)/N]\,\overline{W_m}$*:* (a) *original plot;* (b) $\tau_n = 0.1$*;* (c) $\tau_n = 0.01$

Assuming that the extra process dynamics are governed by $\tau_p = 0.5$, figure 5.5 (plot 'a') shows that this actual error transfer is not SPR, while its Nyquist plot is not completely situated in the right half of the plane. If the decomposition method is applied, the error transfer changes to $[(N - T_2)/N]\,\overline{W_m}$. Plots 'b' and 'c' show the Nyquist plot of this modified transfer when $\tau_n = 0.1$ and $\tau_n = 0.01$ respectively. The decomposition moves the Nyquist plot to the first quadrant, making the real part of the error transfer positive over a larger frequency range.

The second decomposition effect lies in the modification of the output disturbance on the error equation. The original disturbance can be found to be:

$$\nu = \frac{-\tau_p s^2}{(s+1)(\tau_p s^2 + s + 1)} r \tag{5.16}$$

and the disturbance after decomposition, if τ_p is exactly known:

$$\nu = \frac{\tau_p s}{(s+1)(s\tau_n + 1)(\tau_p s^2 + s + 1)} r \tag{5.17}$$

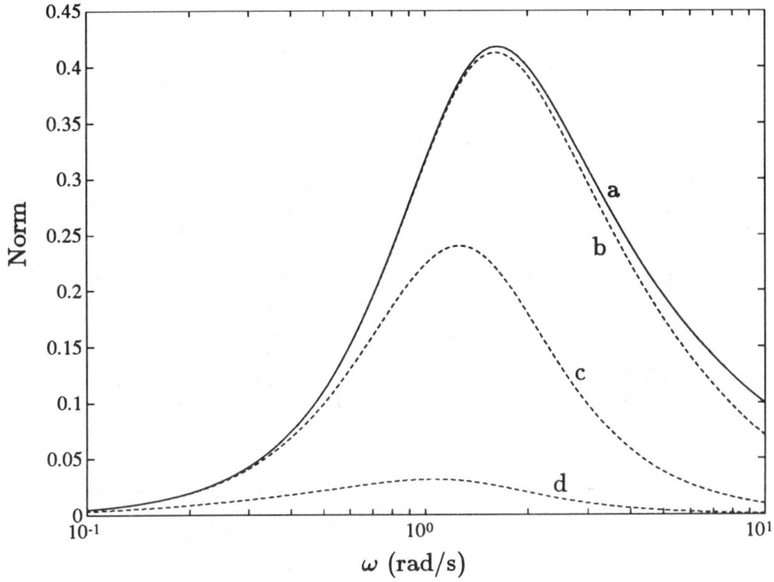

Figure 5.6 First example, disturbance magnitude as a function of frequency: (a) *original disturbance* ($\tau_n = 0$)*;* (b) $\tau_n = 0.1$*;* (c) $\tau_n = 1$*;* (d) $\tau_n = 10$

If decomposition is used, the output disturbance depends on the design parameter τ_n. Figure 5.6 shows the norm of the disturbance in the original case ('a') as a function of the frequency. Plots 'b', 'c' and 'd' show the disturbance magnitude for $\tau_n = 0.1$, $\tau_n = 1$, and $\tau_n = 10$ respectively. It can be seen that a large τ_n is favourable for a small disturbance magnitude. This contradicts the requirement on τ_n for obtaining a more positive real error transfer, for which τ_n should be as small as possible. Note that τ_n can be chosen freely while still being able to implement an exact decomposition.

Simulation results

When no decomposition is applied, the result of figure 5.7 is obtained if a block-type reference signal is applied. The lower-order controller cannot handle the extra dynamics and a proper convergence is not achieved. When the decomposition method is implemented and $\tau_p = 0.5$ is known exactly, the response improves considerably, as shown in figure 5.8. Comparing these figures, it can be seen that the higher-order dynamics are much better handled if decomposition is used,

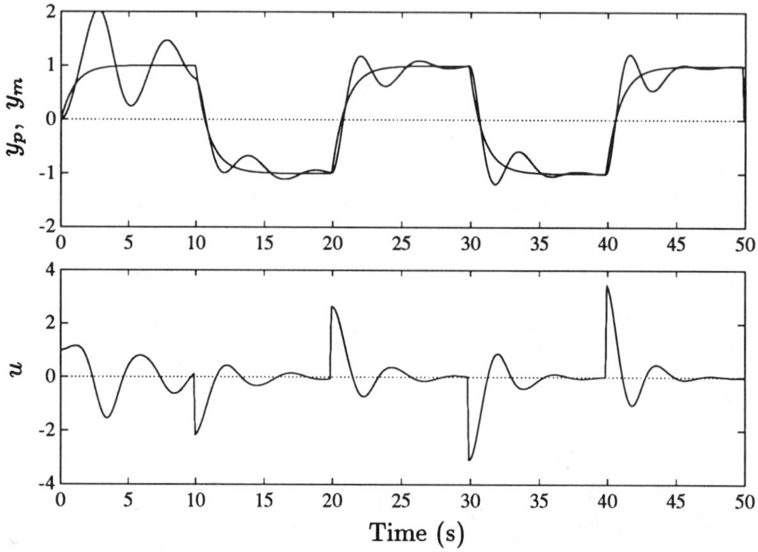

Figure 5.7 First example, result obtained without decomposition

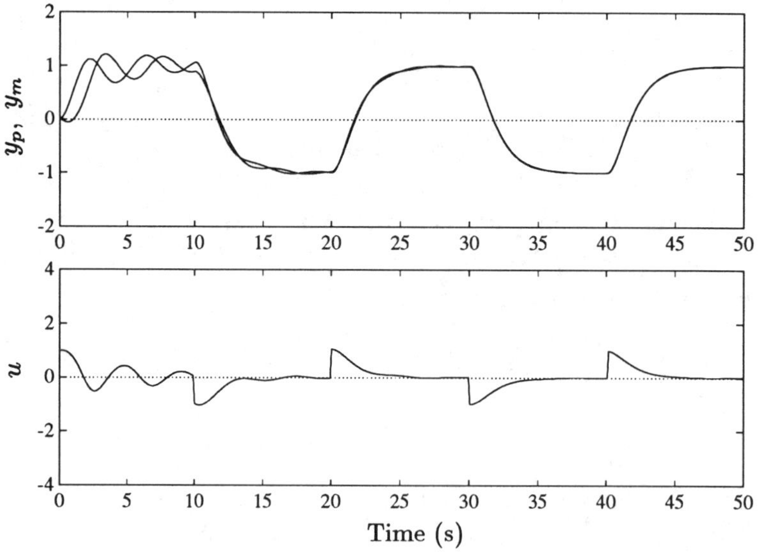

Figure 5.8 First example, result obtained by using exact decomposition

at the cost of somewhat slower behaviour. Note also the considerably reduced control effort. In choosing τ_n, making the real part of the error equation more positive appears to have the largest effect, and τ_n was chosen 0.25. If $\tau_p = 0.25$ instead of $\tau_p = 0.5$, the decomposition is no longer exact. The results, however, are still satisfactory, as shown in figure 5.9.

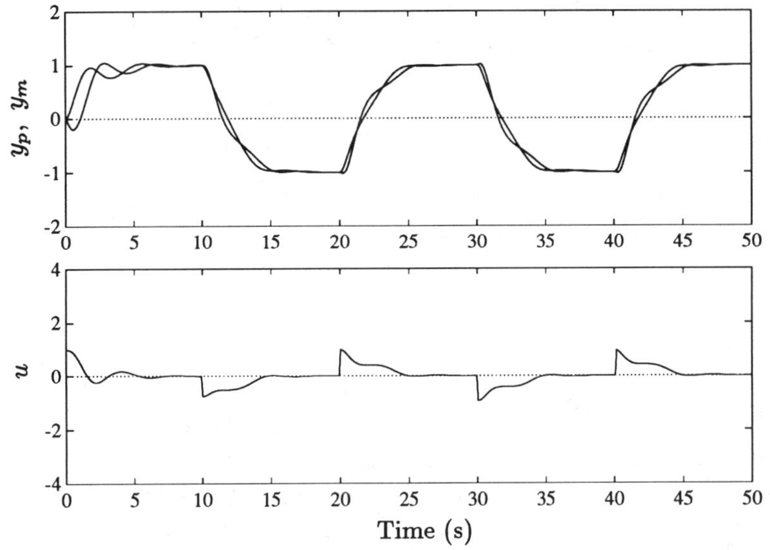

Figure 5.9 First example, result obtained by using non-nominal decomposition

5.3.5 A second example

In this section, an example shows the decomposition method when applied to a process with 'unmodelled' oscillations.

Process and model transfers and decomposition

The nominal process considered has a transfer function:

$$W_p = \frac{k_p}{s\tau_p + 1}$$

The real process, however, is in the form:

$$\overline{W}_p = \frac{k_p}{s\tau_p + 1}\left(\frac{\omega_p^2}{s^2 + 2z_p\omega_p s + \omega_p^2}\right)$$

The MRAC system is designed only for the 'modelled' part $k_p/(s\tau_p+1)$ and thus is based on a system order of 1. The reference model is chosen as:

$$W_m = \frac{b}{s+a}$$

For the decomposition, a form is chosen which resembles the unmodelled part of the process:

$$\frac{T_1}{N - T_2} = \frac{\omega_m^2}{s^2 + 2z_m\omega_m s + \omega_m^2}$$

Note that in this structure, the decomposition polynomials T_1 and $N - T_2$ are determined by parameters z_m and ω_m, which represent design parameters that have an obvious meaning. If z_p is known to be small and hence the process poles are badly damped, it still appears to be useful to choose a proper decomposition model damping ratio z_m to obtain a nonoscillating model output. This illustrates well the borderline between what would be needed to obtain the best results in terms of improvements of the error equation, and what is actually desired. Choosing a low (but accurate) damping ratio z_m decreases the sensitivity of the output error to the unmodelled dynamics, but would result in oscillating behaviour of the reference model output. Hence, an exact decomposition would not yield a desired response. However, in the case of the natural frequency ω_m, the situation is different. Imposing an arbitrary frequency on the oscillating system is not useful and would take much control effort. Therefore, an attempt is made to render the error equation insensitive to the oscillation frequency, by choosing ω_m equal to the nominal value of ω_p (which, of course, may not be known accurately). The polynomial N can be chosen arbitrarily as long as the order of N equals at least 2. Otherwise, the transfer T_2/N has more zeros than poles, introducing implementation problems.

Writing N as:

$$N = s^2 + 2z_n\omega_n s + \omega_n^2$$

it is found that:

$$T_2 = 2\left(z_n\omega_n - z_m\omega_m\right)s + \left(\omega_n^2 - \omega_m^2\right) \tag{5.18}$$

Examining equation (5.18), it can be seen that if $\omega_n = \omega_m$ and $z_n = z_m$, T_2 vanishes and no correction to y_m is made. In that case, the knowledge of the unmodelled dynamics is only incorporated in the reference model. Recalling that the error equation profits from a phase-lead correction to achieve a more positive real part, N can best be chosen relatively fast in this respect, and ω_n is therefore chosen larger than ω_m. The damping ratio z_n is chosen to be 1.

Analysis of the error equation

In order to test the decomposition method, the following process parameters are assumed: $k_p = 0.5$, $\tau_p = 1$, $\omega_p^2 = 4$ and $z_p = 0.7$. The reference model parameters are: $a = 4$ and $b = 4$. The same primary controller as that in the first example is used:

$$u = k_0 r + d_0 y_p$$

Because the nominal process is of order 1, no ASGs are needed and an error-augmenting network is not required. For the undisturbed process W_p, parameter values $k_0 = 8$ and $d_0 = -6$ make $W_p(\boldsymbol{\theta}^{*T}\boldsymbol{\omega}^*)$ equal to $W_m r$. In the disturbed case, the linear part of the error equation equals:

$$\overline{W_m} = \frac{16}{s^3 + 3.8s^2 + 6.8s + 16}$$

Figure 5.10 shows the Nyquist plot of $\overline{W_m}$ (denoted by 'a'), which lies mainly in the second quadrant. Also in figure 5.10, Nyquist plots of $[(N - T_2)/N]\,\overline{W_m}$ are shown with $N - T_2 = s^2 + 4s + 4$, $N = s^2 + 20s + 100$ and $N = s^2 + 63s + 1000$ (plots 'b' and 'c', respectively). To be able to compare the position of these plots in the complex plane, plots 'b' and 'c' are corrected for their lower DC gain. Obviously, making $1/N$ faster has the effect of $[(N - T_2)/N]\,\overline{W_m}$ moving to the first quadrant, increasing the frequency range over which the real part of the error equation's frequency response is positive.

For the disturbance on the output error, an analysis similar to that given in the first example can be made. In this case also, a small ω_n decreases the disturbance the most. A proper compromise value for ω_n must be found such that the overall result is the best.

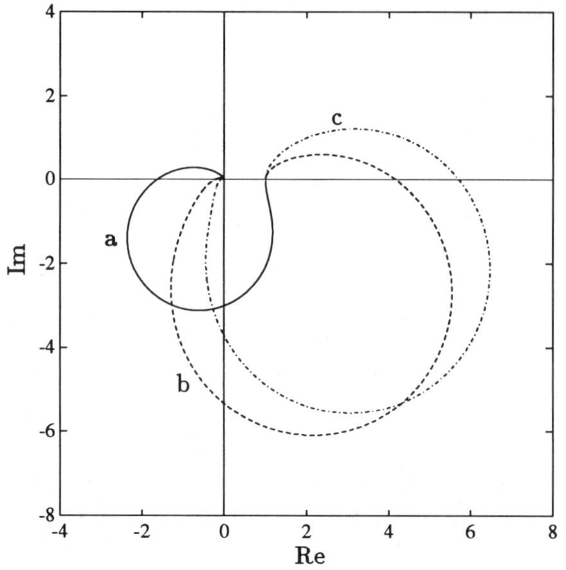

Figure 5.10 Second example, Nyquist plots of $\overline{W_m}$ *and* $[(N - T_2)/N]\,\overline{W_m}$ *with* $N - T_2 = s^2 + 4s + 4$*:* (a) *original plot;* (b) $N = s^2 + 20s + 100$*;* (c) $N = s^2 + 63s + 1000$

Simulation results

If no decomposition is used, the unmodelled dynamics induce unstable behaviour, as seen in figure 5.11. Here, the nominal process parameter values are assumed. If the nominal value of ω_p^2 is known, application of decomposition with $z_m = 1$ and $\omega_m = \omega_p$ gives the result shown in figure 5.12. Here, the polynomial $N = s^2 + 8s + 16$ is found to be an acceptable compromise between the two requirements. If the parameters of the structured unmodelled dynamics are not known exactly, a less exact decomposition arises. To test the robustness of the system with respect to mismatch between process and decomposition parameters, figure 5.13 shows the results with $\omega_p^2 = 8$ and $z_p = 0.25$. It can be seen that the response is still satisfactory.

It should be noted that when using a full-order controller for the process including unmodelled dynamics, perhaps a better response than that shown in figure 5.11 could have been obtained. However, a total number of six parameters need to be adjusted and so the controller is much more complex. In a practical application, an as small as possible number of adjustable parameters is favoured.

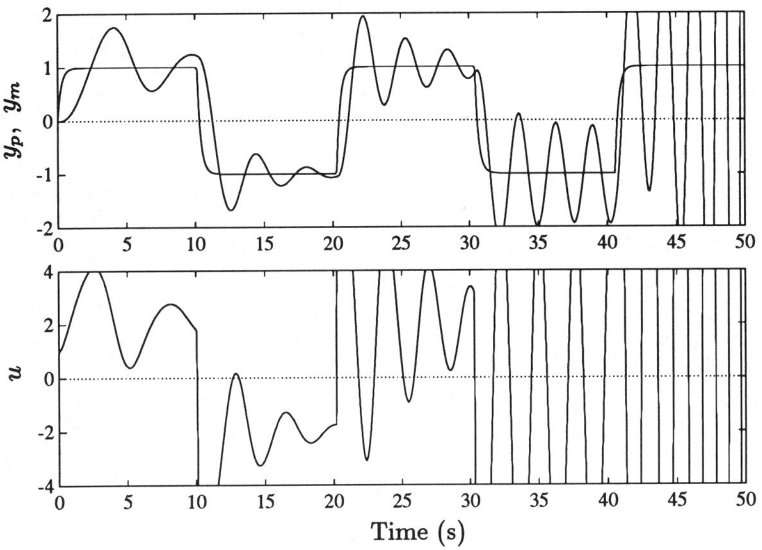

Figure 5.11 Second example, result obtained without decomposition

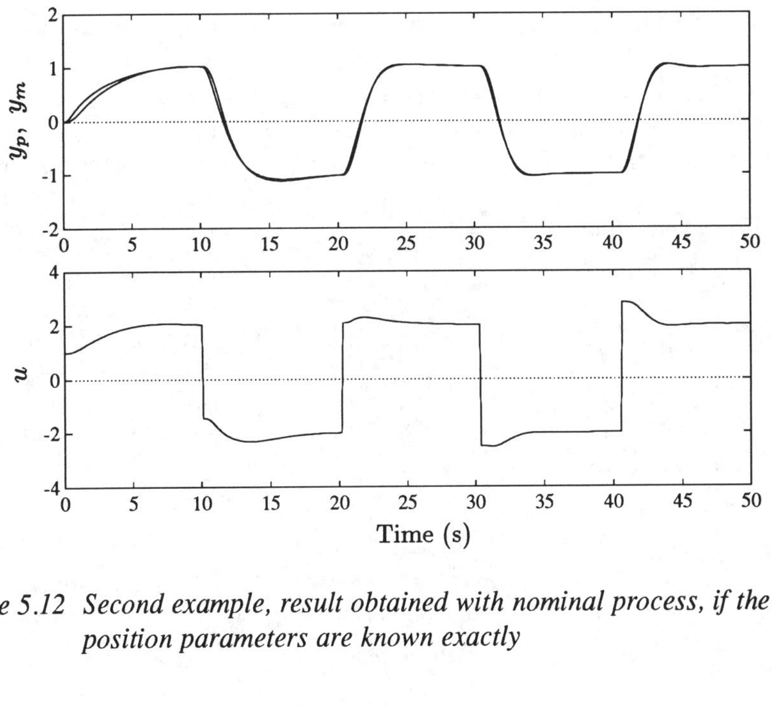

Figure 5.12 Second example, result obtained with nominal process, if the decomposition parameters are known exactly

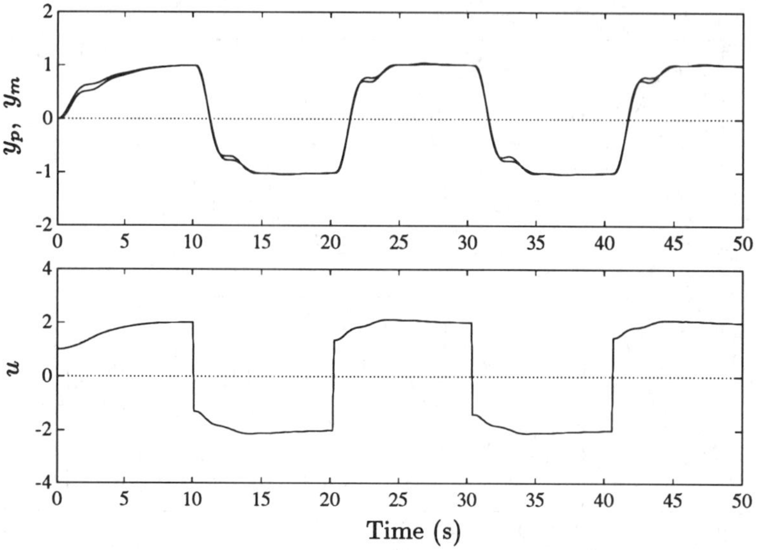

Figure 5.13 Second example, result obtained with non-nominal process

In addition, a full-order controller would require a much larger control effort.

Comparing the responses with decomposition and those without, it can be seen that the decomposition method improves the adaptive system behaviour if unmodelled dynamics with a known structure are present. This improvement is achieved by adjustment of the reference model output, and comparing the original y_m in figure 5.11 with the adjusted y_m in figure 5.12 shows a somewhat slower response because of the decomposition. Hence, the improvement is achieved by adjusting the *control goal* to compensate for a lower controller order than necessary for perfect model matching.

5.3.6 Decomposition design steps

This section summarizes the design steps that have to be taken when applying the decomposition method.

- The first step in the design is the decision as to which part of the process dynamics is to be regarded as 'nominal' and which part 'unmodelled'. The controller order can then be chosen such that the nominal part can be controlled properly and the controller design can take place. The decision can be based on the desired system specifications and the feasibility of a higher-order controller.

- In the next step, the structure of the 'unmodelled' process dynamics is analyzed and copied into the decomposition model. In this model, the parameters are chosen as closely as possible to resemble the actual process parameters. However, the response of the decomposition model is modified such that it represents an acceptable output behaviour. For example, a damping ratio is added to an oscillatory part.
- The modified structure is then transformed into polynomials T_1 and $N - T_2$. In the decomposition structure, the polynomial N can be chosen freely (therewith determining T_2). As shown above, the two decomposition effects impose different requirements on the choice of these polynomials. On the one hand, the roots of N should be chosen as negative as possible to increase the frequency range for which $\mathrm{Re}[\overline{W_m}L(j\omega)] > 0$. On the other hand, to decrease the output disturbance on the error equation, smaller negative real parts of N can be favourable. In addition, the resulting closed-loop behaviour depends on N, and hence N influences the performance of the system. A satisfactory polynomial N for a specific situation can best be determined in simulation.

5.4 Application to a gantry crane scale model

This section describes the application of the decomposition method to the design of a model reference adaptive controller for a scale model of a gantry crane. After a description of the experimental setup and the accompanying mathematical model, the adaptive controller with model decomposition will be presented. Practical results under varying operating conditions are shown which illustrate the useability of the method.

5.4.1 Process description

The experimental setup consists of a scale model of a gantry crane which is shown schematically in figure 5.14. A close-up picture of the trolley and a part of the setup is shown in figure 5.15.

The crane's larger cousin is used in many harbours for moving containers

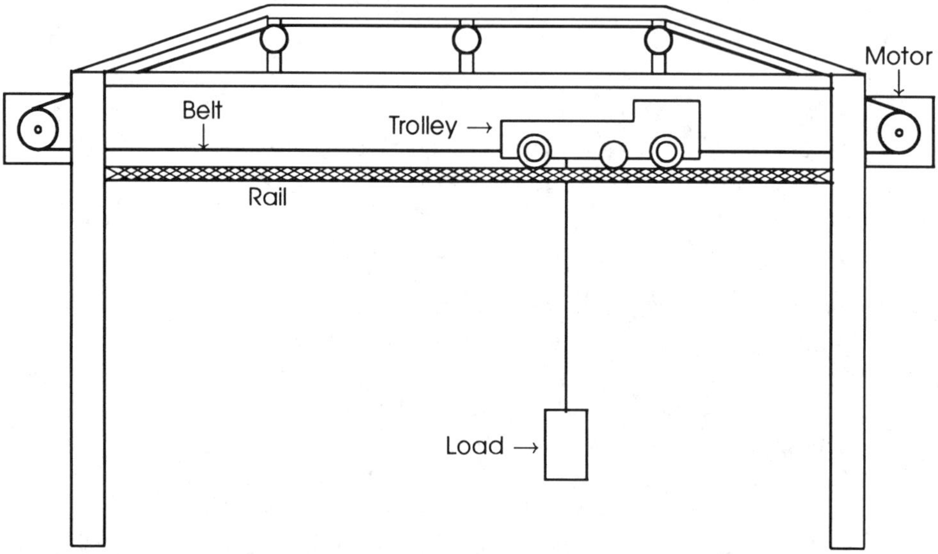

Figure 5.14 Schematic drawing of the gantry crane scale model

and other loads into and out of ships. The crane, as sketched in figure 5.14, is placed on rails parallel to the harbour, so that a load movement in three directions is possible. The crane operator is seated inside the trolley, which moves over the rails spanning the working area. He is responsible for moving the load such that it arrives at the desired spot as quickly as possible, under the constraint that only a very limited overshoot of the load in the horizontal direction is allowed. Violating this requirement may easily result in damaging the load and, for example, the lorry it must be placed on. A real crane may allow a trolley movement of 80 to 90 m, demanding a position accuracy of 0.2 to 0.8 m, depending on the desired load location (Ten Hengel, 1979). After proper training and much practical experience the crane operator gains enough expertise to handle this task. Even the varying swing frequency of the load due to the varying cord length imposes no problems for an experienced operator.

The laboratory scale model has two basic limitations with regard to a real gantry crane. First, all physical dimensions are scaled down dramatically in order to make the experimental setup fit within a laboratory environment. Second, a movement is only possible in two directions: the trolley can move horizontally over the spanning rails, and a motor in the crab can lift the load vertically.

The electrical drives responsible for the trolley movement are fitted in the gantry and move the trolley by means of a flexible steel belt. This belt must be kept

Figure 5.15 Close-up photograph of the trolley and a part of the crane

tense, for which special provisions are made. Despite the limited elasticity of the belt, the remaining flexibility induces some oscillating behaviour, which depends on the actual trolley position. To measure this position accurately, independent of disturbances caused by the belt, a potentiometer is fitted in the trolley. This position sensor is connected to a wheel riding over the rails which has no other function. Any slipping between this wheel and the rail induces measurement offset when the crane has been in operation for a long time.

The maximum trolley range is 2.40 m. In the vertical direction, the lift motor in the trolley can move the load over a range of 0.66 m (the cord length can vary between 0.63 m and 1.29 m). The maximum load mass is 10 kg, and to keep the cord straightened a minimum load of 0.5 kg is required.

The control goal is a fast and accurate movement of the load in the horizontal direction, independent of the cord length. During the movement, the cord length is assumed to be fixed, which corresponds with the situation in practice: all load lifting and lowering is done when the trolley is in a fixed position. The swing frequency ω of the load depends only on the actual cord length l:

$$\omega = \sqrt{g/l}, \qquad g = 9.81 \text{ N/kg}$$

Therefore, ω varies between 2.76 and 3.95 rad/s. In addition, as will be shown later, the swing amplitude depends on the cord length. The control signal is the input voltage u at the trolley motor. For the controller, three sensors are available. The position sensor on the trolley, as mentioned above, provides a measurement of the trolley position x_t. Other potentiometers in the trolley measure the angle α between the cord and the vertical axis, and the cord length l (figure 5.16). By combining the information obtained from these sensors, the actual horizontal load position x_l can be calculated:

$$x_l = x_t + l \sin(\alpha) \tag{5.19}$$

Only in this position is x_l used as feedback information for the controller, so that the crane is considered as a single-input, single-output system. The mathematical model of the crane will be derived in the following section.

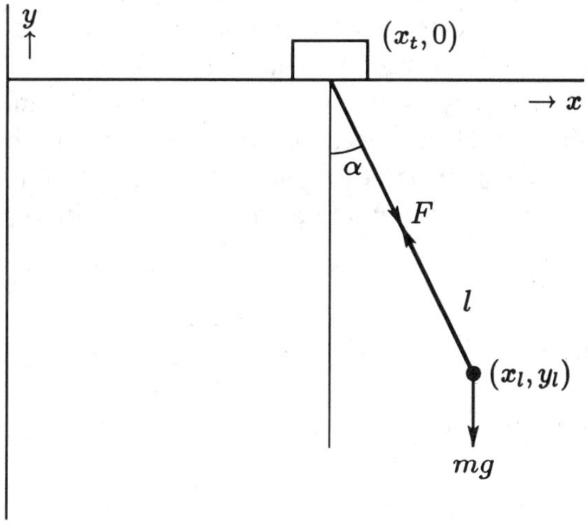

Figure 5.16 Definition of the coordinate system and the various variables which play a role in the mathematical model of the crane; g is gravity constant; F is the cord tension; m is load mass

5.4.2 Mathematical model of the crane

Neglecting the electrical time constant in the motor, the voltage u on the trolley motor results in a trolley position x_t given by:

$$x_t = \frac{k}{s(s\tau + 1)} u$$

Flexibility in the transmission belt is neglected because its influence is small compared with that of the load swing. This flexibility has a frequency of 35 rad/s, which is about ten times higher than the swing frequency, and has a relatively small amplitude. However, if the control goal is an accurate trolley position instead of an accurate load position, this flexibility should be included.

The time constant τ depends on the motor parameters (back-EMF constant), on the trolley inertia, and on viscous friction between the trolley and the rail. Measurements have yielded values of $k = 0.35$ and $\tau = 0.065$ s. Due to Coulomb friction, the motor does not come into movement if the voltage is below a certain threshold. This effect does not depend on the trolley position, but is asymmetric: the negative and positive thresholds are -1.6 V and $+1.2$ V, respectively. Because these values are more or less constant, a direct compensation is implemented (with a small safety margin, so that a zero controller output never results in a trolley movement). In the transfer function x_t/u, it is assumed that the swinging load does not influence the trolley position. It will be shown later that this assumption is valid and that this disturbance need not be incorporated in the model.

Derivation of the equations of motion of the crane completes the modelling part. Taking the x, y-plane as in figure 5.16, the load position (x_l, y_l) is:

$$\begin{aligned} x_l &= x_t + l \sin(\alpha) \\ y_l &= -l \cos(\alpha) \end{aligned} \tag{5.20}$$

Applying Newton's law yields:

$$\begin{aligned} m\ddot{x}_l &= -F \sin(\alpha) \\ m\ddot{y}_l &= F \cos(\alpha) - mg \end{aligned} \tag{5.21}$$

in which m is the mass of the load, F is the tension in the cord and g is the gravity constant. Eliminating F from equation (5.21) gives:

$$\ddot{x}_l \cos(\alpha) + \ddot{y}_l \sin(\alpha) = -g \sin(\alpha) \tag{5.22}$$

Differentiating equation (5.20) twice, while assuming that l is constant, yields:

$$\begin{aligned} \ddot{x}_l &= \ddot{x}_t - l\dot{\alpha}^2 \sin(\alpha) + l\ddot{\alpha} \cos(\alpha) \\ \ddot{y}_l &= l\dot{\alpha}^2 \cos(\alpha) + l\ddot{\alpha} \sin(\alpha) \end{aligned} \tag{5.23}$$

Substituting equation (5.23) in equation (5.22) gives:

$$\ddot{x}_t \cos(\alpha) + l\ddot{\alpha} = -g \sin(\alpha)$$

or:

$$\ddot{\alpha} = -\frac{g \sin(\alpha) + \ddot{x}_t \cos(\alpha)}{l} \tag{5.24}$$

Equation (5.24) gives a relation between the trolley acceleration $\ddot{x}_t$ and the angular acceleration $\ddot{\alpha}$. Assuming that α has a small enough value to allow an approximation $\sin(\alpha) \approx \alpha$ and $\cos(\alpha) \approx 1$, the transfer from $\ddot{x}_t$ to α becomes:

$$\alpha = \frac{1/l}{s^2 + g/l}\ddot{x}_t \tag{5.25}$$

Combining equation (5.25) with equation (5.19) yields:

$$\frac{x_l}{u} = \frac{k}{s(s\tau + 1)}\left(1 + \frac{s^2}{s^2 + g/l}\right) \tag{5.26}$$

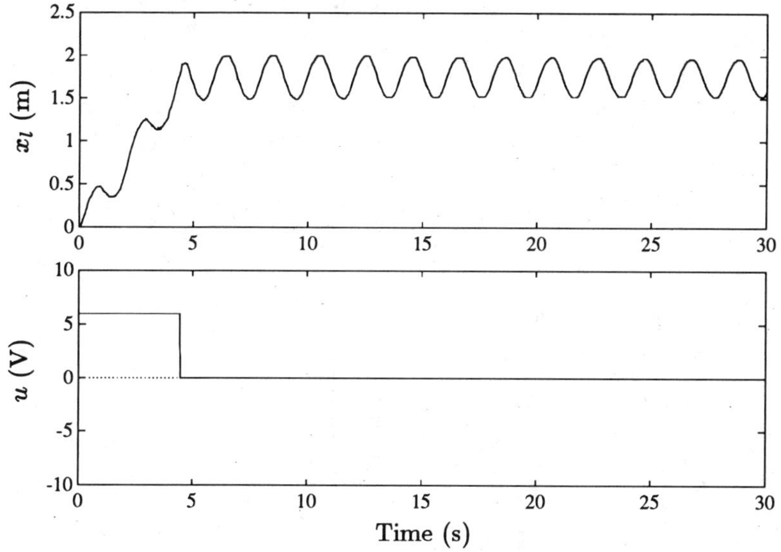

Figure 5.17 Open-loop crane response if a 6 *V pulse input is applied*

Figure 5.17 shows the open-loop response of the crane if a 6 V pulse input is applied. The oscillating behaviour is clear. Equation (5.26) describes the simplified transfer from input voltage u to horizontal load position x_l. Coulomb friction, belt flexibility and the goniometric relations have all been neglected, as is the influence of the load swing on the trolley. This latter effect will now be analyzed to show the validity of this simplification.

For the trolley, which weighs 9.7 kg, a 'virtual mass' can be defined as the relation between an externally applied force on the trolley and the resulting acceleration. Because of the trolley motor, which acts as a 'dynamic brake', the virtual mass is much larger than the actual mass. By applying a force on the trolley and studying the trolley movement, the virtual mass can be determined. This approach leads to a virtual mass of $M = 170$ kg, which is large with respect to the load mass. The ratio m/M expresses the influence the load has on the trolley movement. The allowed range of the load mass results in a value of m/M between 0.003 and 0.06, which is sufficiently small to allow the neglect of this disturbance on the trolley.

5.4.3 The adaptive controller

Applying an MRAC scheme without modification to the gantry crane would necessitate adjustment of eight parameters (Butler *et al.*, 1991a). In practice, such a system shows unacceptably nonconverging behaviour. This is mainly caused by the ASGs, which generate filtered derivatives of y_p, amplifying measurement noise which disturbs the adaptation. In addition, some Coulomb friction in the sensor for the angle α introduces extra high-frequency disturbances in the measurements, and so does the neglected high-frequency belt flexibility. Therefore, the controller will be designed only for the 'nominal' part, which is chosen as:

$$W_p = \frac{k}{s(s\tau + 1)}$$

The reference model is chosen as:

$$W_m = \frac{b_0}{s^2 + a_1 s + a_2}$$

The control signal is made up of four elements:

$$u = k_0 r + \frac{c_1}{s+n}\, u + \frac{d_1 s}{s+n}\, y_p + d_0 y_p \tag{5.27}$$

The auxiliary signal generator polynomials are represented by the factors $s+n$. In principle, n can be chosen arbitrarily because the reference model does not have any zeros. Because the controller is designed for the case where oscillations are not present, filters L^{-1} of order 1 must be chosen such that $W_m L$ is SPR. The reference model parameters are chosen as $b_0 = 20$, $a_1 = 9$ and $a_2 = 20$, yielding poles at $s = -4$ and $s = -5$. L is selected as $s + 5$, making $W_m L = 20/(s + 4)$.

In order to be able to deal with the 'unmodelled' oscillations, the decomposition method is applied. For the decomposition model, a form is selected which resembles the 'unmodelled' process part. According to the guidelines given in section 5.3.6, a damping ratio z_m is added to guarantee a desired model output:

$$\frac{T_1}{N - T_2} = \left(1 + \frac{s^2}{s^2 + 2z_m\omega_m s + \omega_m^2}\right)$$
$$= \frac{2s^2 + 2z_m\omega_m s + \omega_m^2}{s^2 + 2z_m\omega_m s + \omega_m^2}$$

As in the second example in section 5.3.5, the decomposition polynomials T_1 and $N - T_2$ are determined by parameters z_m and ω_m. N can be chosen arbitrarily, although it is assumed that the order of N equals 2. Writing N as:

$$N = s^2 + 2z_n\omega_n s + \omega_n^2$$

it follows that:

$$T_2 = 2\left(z_n\omega_n - z_m\omega_m\right)s + \left(\omega_n^2 - \omega_m^2\right)$$

The natural frequency of the model ω_m^2 is chosen equal to the minimum value of g/l, which is 7.6. Practical experiments showed the best results when using this setting, because the decomposition is then tuned for the largest amplitude of the swing. To ensure proper model damping, z_m is chosen as 1. The parameters z_n and ω_n^2 are chosen as 1 and 30 respectively. The value of $\omega_n^2 = 30$ is large enough to provide a considerable phase lead, and small enough to allow a computer implementation at a sample time of 30 ms (see the following section).

5.4.4 Practical results

The gantry crane sensors and motors are connected to a hardware interface, which conditions all signals to voltages between −10 and +10 V. The interface is connected, via A/D and D/A converters, to a VAXstation II minicomputer, which is equipped with the real-time shell MUSIC (MUlti-purpose SImulation and Control). The controller is written in `FORTRAN` and interfaces to a software module which operates the A/D and D/A converters. Further, this module calculates the actual load position x_l from the measurements, and removes the sensor offsets. The computing effort required for the complete system, including the adaptive controller, allows a sample time of 30 ms.

Applying the augmented error method as presented above, either with or without decomposition, results in badly damped behaviour. This effect appears to be caused by the number of adjustable parameters, in this case four, which result in too much sensitivity for all disregarded nonlinearities and belt flexibility. Decreasing the number of parameters to two:

$$u = k_0 r + d_0 y_p$$

increases the system performance considerably. This simplification is allowed because the mechanical time constant $\tau = 0.065$ s is relatively small compared to the reference model time constants and can therefore be ignored. Figure 5.18 shows the response using the original adaptive scheme without decomposition, for $l = 1.00$ m. Note that there is a remaining overshoot of 15% because of

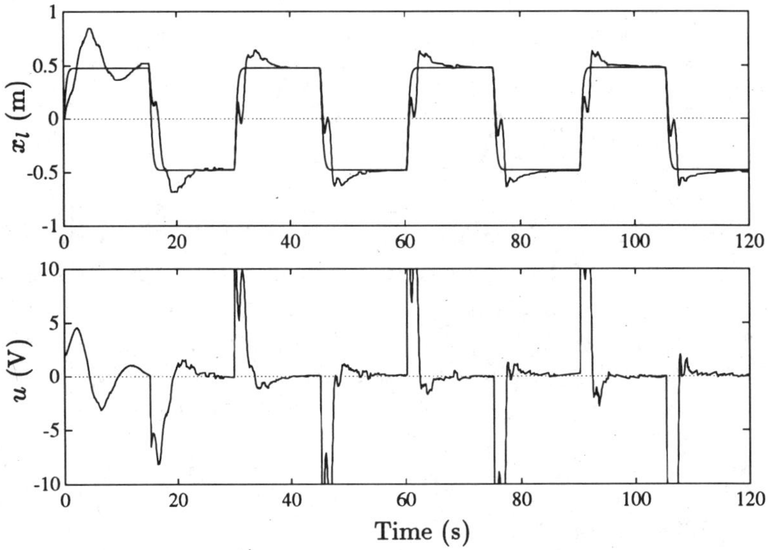

Figure 5.18 Crane response to a block-type input with a 45 cm amplitude without using decomposition, $l = 1.00$ m

the inability of the controller to deal with the swing dynamics. Practically, this means that during the transient a large error signal occurs due to the inevitable load swing. This makes the adaptation increase the feedforward parameter k_0. When y_m has settled, the large value of k_0 results in overshoot, and the adaptive system starts decreasing k_0 to obtain a correct DC gain. This pattern repeats itself at every set-point change.

Figure 5.19 shows results using the same parameters, but with the decomposition method implemented. Figures 5.20 and 5.21 show the crane behaviour

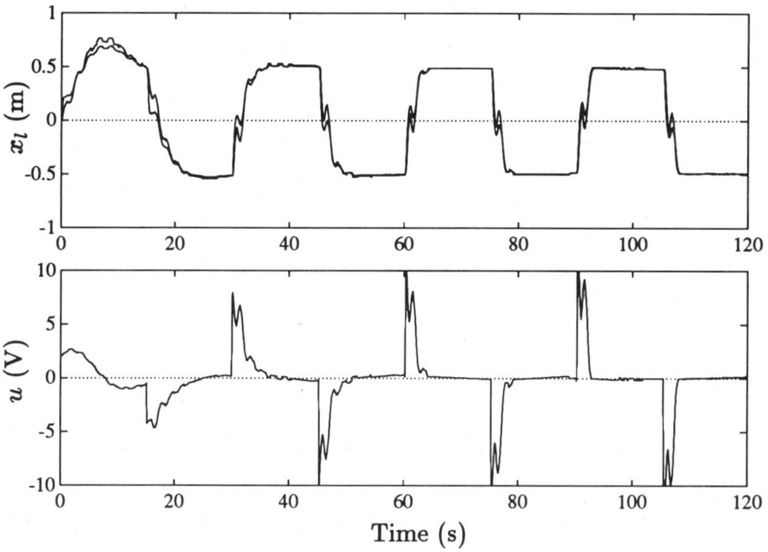

Figure 5.19 Crane response to a block-type input with a 45 *cm amplitude obtained by using decomposition,* $l = 1.00$ *m*

with $l = 0.63$ m and $l = 1.29$ m respectively. It can be seen that the overshoot is decreased considerably, and that the decomposition method is robust with regard to the cord length. The diagrams show that this improvement is achieved by the changing reference model, which depends on the actual process output in such a way that the process-model error is made less sensitive to the load swing. In effect, the model exhibits a similar 'swing' as the process, avoiding a large error signal during the transients. Hence, the practical useability of the decomposition method is demonstrated.

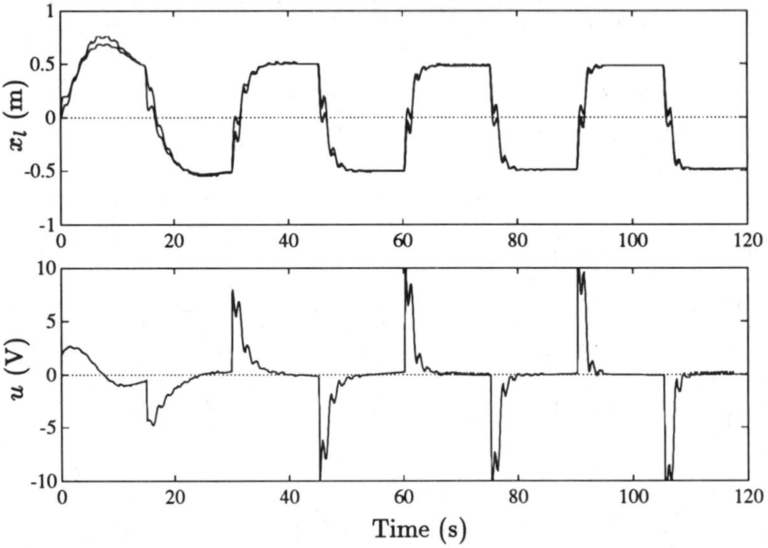

Figure 5.20 Crane response to a block-type input with a 45 *cm amplitude obtained by using decomposition,* $l = 0.63$ *m*

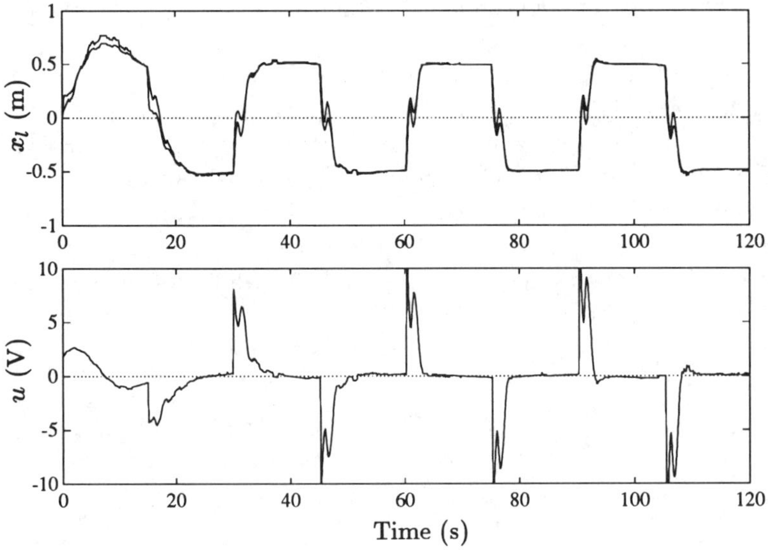

Figure 5.21 Crane response to a block-type input with a 45 *cm amplitude obtained by using decomposition,* $l = 1.29$ *m*

5.5 Summary

The problem of unmodelled dynamics in the process differs from that of external disturbances, mainly because unmodelled dynamics (or state-dependent disturbances) cannot be guaranteed to be bounded. Unmodelled dynamics may easily induce instability in the adaptive loop. One mechanism that plays a role is *slow drift instability*, which is caused by a combination of too low a degree of persistent excitation of the reference signal and the presence of some measurement noise. The noise causes parameter drift, which may lead to instability due to the unmodelled dynamics. Another cause of instability is the attempt to match the model perfectly in a frequency range in which the unmodelled dynamics are active. In this frequency range, perfect model matching is not possible due to the primary controller being designed only for the nominal process.

In terms of the error equation, unmodelled dynamics cause two problems. First, the SPR property of the linear part is violated. Second, a nonzero disturbance on the error equation arises, symbolizing the inability of the primary controller to achieve perfect model matching. If the unmodelled dynamics have a relatively small influence on the process output, the same adaptive law modifications can be applied as in the case of external disturbances. However, if the unmodelled dynamics severly distort the process output and lie in the bandwidth of the nominal part of the process, some other solutions such as reference model decomposition, should be applied.

Reference model decomposition can improve the system behaviour if the structure of the unmodelled dynamics is known. In addition, some knowledge of the parameters in the unmodelled part should be gained. The process dynamics are divided into three classes: 'nominal', 'structured unmodelled', and 'unmodelled'. The primary controller is equipped for the nominal part. Reference model decomposition makes the adaptive system robust with respect to the structured unmodelled part. The real unmodelled dynamics (usually of high frequency) can be dealt with by modifications of the adaptive law, which guarantee stability if the unmodelled dynamics play a minor role in the process output.

Decomposition can be regarded as a method that allows the use of a lower-order controller than is required for perfect model matching. This is useful in a practical application, in which the requirement that rough knowledge about the unmodelled dynamics structure and parameters must be present is generally satisfied. Using a lower-order adaptive controller without decomposition, the adaptive system tries to impose the reference model response on the process, without taking into consideration the limited model-matching possibilities. By not taking into account the limited controller capabilities which are due to the unmodelled dynamics, instability may easily occur.

The decomposition method can be considered as yielding a compromise between perfect model matching and stability. The reference model is adjusted to the process such that its output takes the actually achievable process response into account. This adjustment is of a *direct* form. An adaptive counterpart will be presented in chapter 6. Because the reference model output is adjusted for the unmodelled dynamics, these dynamics have less influence on the output error and thus on the parameter adaptation. The model adjustment results in a somewhat slower reference model output. However, the stability properties are increased considerably, both theoretically and in practice. One side effect is reduced control effort.

The decomposition has two effects on the error equation of the adaptive system. First, the real part of the error equation's frequency response is made positive over a wider frequency range. Second, the output disturbance on the error equation can be reduced. These two effects result in improved system properties in the sense of the existing robustness analysis methods.

Problems

5.1 Explain why attempting perfect model matching in a frequency range where the unmodelled dynamics play an important role may cause instability. In this respect, would you consider a 'fast' reference model more dangerous than a 'slow' one if the unmodelled dynamics are important somewhere outside the process bandwidth?

5.2 Using the following nominal process transfer W_p, full-order process $\overline{W_p}$ and reference model W_m:

$$\begin{aligned} W_p &= \frac{1}{s+1} \\ \overline{W_p} &= \frac{10}{(s+1)(s+10)} \\ \Longrightarrow \widetilde{W_p} &= \frac{10}{(s+1)(s+10)} - \frac{1}{s+1} = \frac{-s}{(s+1)(s+10)} \\ W_m &= \frac{4}{s+4}, \end{aligned}$$

calculate $\overline{W_m}$ and $\widetilde{W_m}$. Use a control $u = k_0 r + d_0 y_p$.

5.3 Assuming you know W_p and $\overline{W_p}$ exactly, propose a possible decomposition of the unmodelled part for the above exercise.

5.4 For the same exercise, give the original output disturbance on the error equation. Calculate also the disturbance obtained when using a decomposition such that $T_2 = 0$.

6

Adaptive Model Adjustment: an Application

If the adaptive control goal of making the process follow the reference model exactly can for some reason not be achieved, on-line adjustment of the model may prove valuable. In chapter 5, this strategy was used for the case of structured unmodelled dynamics, which make it impossible for the primary controller to achieve perfect model matching. A similar problem setting occurs when the *process* capabilities are not constant, but depend on one or more of its parameters. For example, the maximum speed with which a robot can move depends on the weight of the load, which may be unknown. In such a case model adjustment can also be valuable, as will be made clear in this chapter.

6.1 Introduction

If the process is completely linear and of known order, it can always be made to follow exactly a reference model that has the correct relative degree. In practice, however, every process imposes some limits on its input. For example, a DC motor that is designed for input voltages between -10 and $+10$ V will not survive if the adaptive controller demands a motor response that requires 1000 V on the motor input. Hence, every process's input must be limited between safe values. This implies directly that not every possible linear reference model can be followed exactly by the process. A 'fast' reference model in particular requires wide control boundaries and is therefore more likely to cause input saturation than a 'slow' reference model. To ensure that perfect model following, a prerequisite for stability, is maintained, a fixed reference model should be chosen such that, even in the worst case, the process can actually follow the model.

However, if the process parameters are unknown, the 'fastest' feasible reference model cannot be determined easily. In such a case the reference model should be chosen with even more care. In addition, if the process parameters vary, the maximum obtainable performance also varies. For example, if one has the aim of controlling a DC motor in a time-optimal fashion, a large motor load will decrease the maximum obtainable motor acceleration. A time-optimal reference model should therefore be chosen such that even with the largest possible motor load, inducing the slowest possible motor response, it can always be followed. Hence, the reference model is chosen very 'slow', which is obviously disadvantageous at those times where the motor *can* be driven faster (i.e. when its load is smaller).

This problem setting has some similarities with that of the 'structured unmodelled dynamics' in chapter 5. In both cases, the choice of a fixed reference model that yields both a *desired* and an *obtainable* response is difficult or even impossible. In chapter 5, a fixed reference model could not take into account the limited model-matching capabilities due to the unmodelled dynamics. Instability could easily result from this violation of the perfect model-matching condition. In the present chapter, the design criteria cannot be met if a fixed reference model is used. In chapter 5, the problem was solved by adjusting the reference model directly such that the disturbance did have as small as possible an influence on the output error and hence on the adaptation mechanism. In this chapter, an *adaptive* form of model adjustment will be applied that adjusts the reference model to the actual capabilities of the process. The rest of this chapter is devoted to one particular problem: the time-optimal control of a direct-drive DC motor.

In time-optimal control, the input saturation should be active during most of the control period. Some characteristics of the problem are:

- The unknown parameter (in this case the load inertia) is quickly varying and cannot be regarded as constant in time. In particular, the load inertia is changed during standstill of the motor. As soon as the motor begins to move, the adaptation should immediately become active, such that in the first position change the response already equals that of the model. A gradual adaptation, as we have seen in all examples so far, is insufficient in this case. A fast adaptation type, such as proportional adaptation, is therefore required.

- The system is highly nonlinear due to a speed-dependent current limiter (commutation limitation). The desired time-optimal behaviour demands activation of the limiters during the largest part of the movement, which makes the motor response dependent on the magnitude of the set-point change.

- Because the objective is to obtain a *time-optimal* control for every possible load inertia, the *desired* response depends on the actual motor parameters (the load inertia), which are unknown. This means that a fixed reference model does not suffice for achieving the control goal.

In the example given in this chapter, the reference model is adjusted to the process capabilities through identification of the load inertia. The estimated value of the load inertia is a measure for the maximum obtainable performance and is used in the reference model. This way, the model is made to represent the desired time-optimal motor behaviour (Butler *et al.*, 1989).

6.2 The direct-drive motor

A direct-drive motor is in fact just a very strong DC motor. This section presents some important preliminaries on direct-drive motors that are needed in subsequent sections treating the controller design.

6.2.1 Introduction

Recently, DC motors that deliver a high torque on the motor axis have become available, due to the development of new magnetic materials. The high acceleration torque of these motors enables a direct coupling of the load to the motor axis, which makes a transmission superfluous. The disadvantages of a gear train, such as backlash and friction, can hence be avoided. Further, the use of permanent magnets instead of conventional field windings makes the direct-drive motor compact and its use attractive for robot applications.

However, the absence of a gear train involves considerable sensitivity of the motor behaviour to variations in the load inertia. This becomes clear on inspection of the following equation for the load acceleration α_p, in which n is the gear ratio, M_m is the motor torque, J_l is the load inertia and J_m is the inertia of the motor axis:

$$\alpha_p = \frac{M_m n}{J_l + J_m n^2}$$

Increasing the gear ratio n decreases the load acceleration at a constant motor torque. In addition, the influence of the load inertia J_l in the denominator decreases compared with $J_m n^2$, and so the sensitivity of the load acceleration α_p to variations in J_l becomes smaller. For $n = 1$ the value of α_p depends heavily on J_l. In a practical application, such as a robot, the load inertia may be unknown and can change during operation.

6.2.2 Direct-drive motor model

A model of a DC motor is shown in figure 6.1, where R_m is the copper resistance, L_m is the motor inductance, K_t is the torque constant, K_b is the back-electromotive-force (EMF) constant, J_t is the total inertia ($= J_l + J_m$), f is the viscous friction, and M_w is the Coulomb friction. It can be seen from figure 6.1 that the model contains an electrical transfer function $1/(sL_m + R_m)$ and a mechanical transfer $1/(sJ_t + f)$. Coulomb friction, which is caused by magnetic and

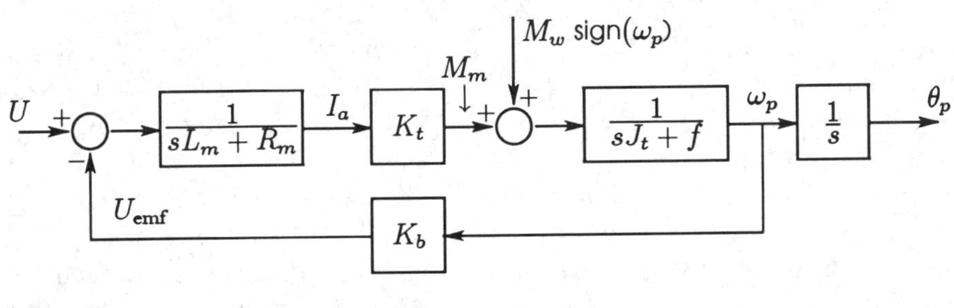

Figure 6.1 Block diagram of DC motor

mechanical hysteresis, results in a torque that has a magnitude M_w and a sign that is opposite to the sign of ω_p. This static model of the Coulomb friction, represented by the factor $M_w \text{sign}(\omega_p)$, is only an approximation of the real friction effects (Walrath, 1984) but is sufficiently accurate for our purpose.

6.2.3 Current control

To protect the motor from overload, the manufacturer places specified constraints on the motor current. Three types of current limitation are necessary:

1. Peak-current limitation. The maximum motor current must be limited to avoid demagnetization of the permanent magnets. For the motor used, the allowed peak current equals 10.4 A.
2. Continuous-current limitation. To prevent the motor from overheating, the maximum constant current must be limited to approximately half the peak current (4.5 A for the motor used). The continuous-current limitation only becomes active after a certain period of time and can be disregarded in position control. However, if torque control is used, longer periods of maximum applied power can occur and in such a case the continuous-current limitation becomes important.
3. Commutation limitation. Sparking under the brushes increases in proportion to both the motor speed and the motor current. To prevent burn in of the brushes, the product of I_a and ω_p must be limited. For the motor used, $I_a \cdot \omega_p$ must be kept below 49 A rad/s.

To be able to limit the motor current I_a, a current control loop is used which has a set point I_s as input and ensures that I_a becomes equal to I_s (figure 6.2). When the current loop operates properly, it is possible to limit I_a by limiting I_s to $I_{\max}$, as shown in figure 6.2. Within the current loop, a servo amplifier with a transfer function $K_{sv}/(s\tau_{sv}+1)^2$ ($K_{sv} = 20$ and $\tau_{sv} = 0.1$ ms) provides the power needed. Its output is limited to ± 100 V, but this limitation appears not to be important where the current limiters are usually activated before the maximum voltage is reached. In addition to the servo amplifier, there is a PI compensator with a transfer function $K_p(1 + 1/s\tau_i)$ ($K_p = 1$ and $\tau_i = 3$ ms), which prevents a steady-state error for step inputs in the transfer from I_s to I_a. In addition, this compensator diminishes the disturbance due to the back-electromotive-force voltage U_{emf} on I_a.

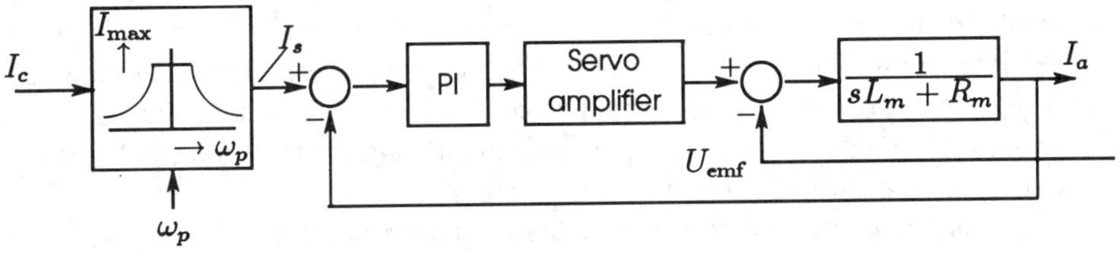

Figure 6.2 Current loop used to allow current set points

The time constant of the current loop is smaller than 1 ms and its step response has no overshoot. The parameters in the current loop are chosen such that the effect of the back-electromotive-force voltage U_{emf} on I_a is small. Because of the high speed of the current loop compared to the time constants in the motor, this loop may be approximated by a transfer function of 1. The current loop enables a specified current to be imposed on the motor, and results in the simplified scheme of figure 6.3.

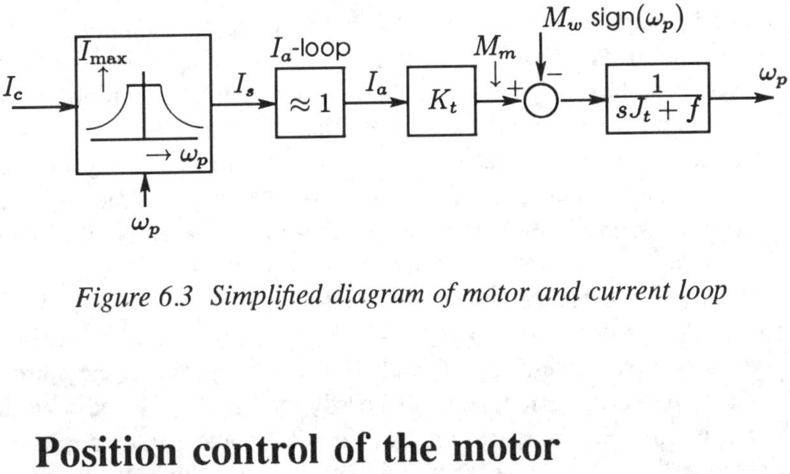

Figure 6.3 Simplified diagram of motor and current loop

6.3 Position control of the motor

With the desired time-optimal control, a step-wise set-point change causes the motor to first accelerate maximally during a certain period of time, followed by a time period of maximum deceleration. The switching moment must be such that at the end of the control the motor angle θ_p equals the set point θ_d. Viscous and Coulomb friction causes the maximum acceleration rate to be smaller than the maximum deceleration rate. A large load inertia J_l requires a 'late' switching moment. However, pure time-optimal control (bang–bang control) has the disadvantage that its disturbance rejection is minimal. In addition, implementation of bang–bang control requires knowledge of the load inertia J_l, while the switching moment depends on this parameter. Hence, the starting point for the design of the position controller will be a conventional PID controller, which will be tuned such that an approximately time-optimal control occurs. Such tuning depends, however, on the load inertia and the set-point magnitude, and hence can be determined only for a particular combination of these two.

Measurement information is available for both the position θ_p and the velocity ω_p. The motor angle θ_p is measured by a *resolver*. The hybrid resolver control transformer (HRCT), interfacing the resolver, has a 14-bit digital set-point input and an input for the resolver signal, and gives the sine of $(\theta_d - \theta_p)$ as output. The resolver principle is based on the fact that the voltage induced in a receiver coil depends on the angle between the transmitter and the receiver coil. By fixing one of the coils to the motor frame and the other to the motor axis, the induced voltage is a measure of θ_p. A 400 Hz sine wave is input to the transmitter and induces an AM-modulated signal in the receiver, which is demodulated in the HRCT by applying asynchronous detection. This causes an 800 Hz disturbance signal in the HRCT output, which is therefore filtered by a second-order filter with a transfer function $1/(s\tau_f + 1)^2$ ($\tau_f = 1.5$ ms), having a disturbance rejection of 35 dB at 800 Hz. An advantage of the resolver over, for example, an incremental encoder, is its high resolution. However, the nonlinear sine function, which is elementary to the resolver, prohibits large set-point changes, while angles larger than $\pi/2$ rad cannot be distinguished from angles smaller than $\pi/2$ rad.

As well as a proportional factor K_p and velocity feedback K_v, an integrating factor K_i is implemented to avoid a steady-state error caused by Coulomb friction. To prevent the integrator from winding up, it only operates when the current limiters are not activated (i.e. when the absolute value of I_a is less than or equal to $I_{\max}$). The full scheme of the motor, the current loop, and the PID position controller is depicted in figure 6.4. Due to the unknown inertia of the load, a fixed set of parameters K_p, K_i and K_v cannot ensure that the closed-loop response meets its design criteria for all possible values of the load inertia and set-point magnitude. If, for a certain load inertia J_l, the parameters provide a suitable response, a larger value of J_l will result in too high an overshoot. A smaller load inertia, however, results in an overdamped, and so non-time-optimal, response. Further, the magnitude of the step input affects the motor behaviour due to the presence of current limiters. Applying model reference adaptive control, it was found in simulations that adjustment of only K_v was sufficient to be able to impose an approximately time-optimal response. The controller is shown in figure 6.4 as an extension of the PID controller.

The adaptive law is of a standard form, as derived in chapter 1, assuming the process is linear and neglecting the electrical time constants, and includes a

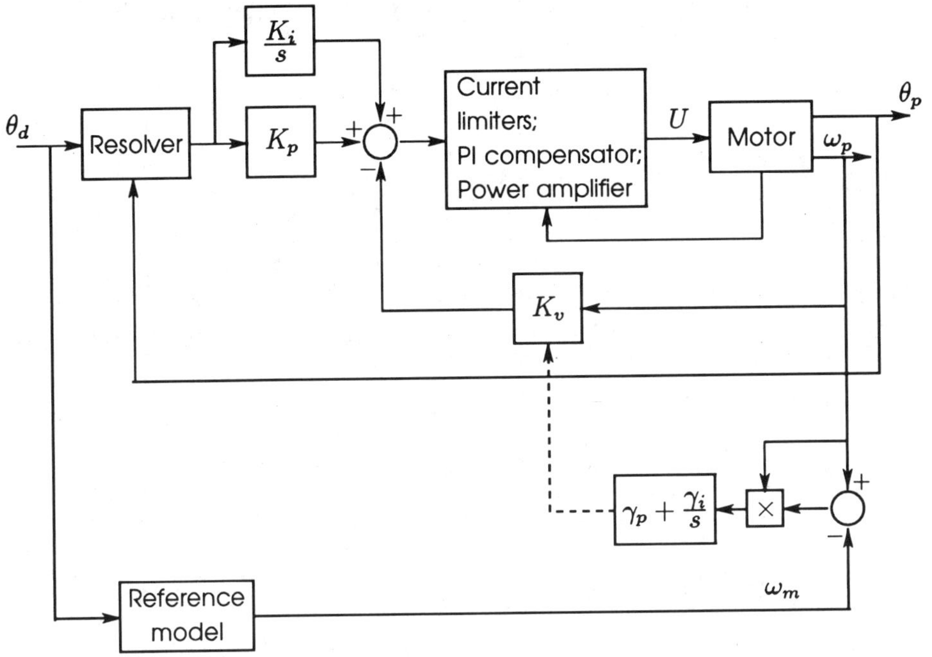

Figure 6.4 Full scheme of direct-drive motor system, with PID controller and MRAC parameter adjustment

proportional term:

$$K_v = \gamma_i \int_{\tau=0}^{t} \omega_p (\omega_p - \omega_m) \, \mathrm{d}\tau + \gamma_p \omega_p (\omega_p - \omega_m)$$

Because only K_v is adjusted, θ_p need not be incorporated in the adaptive laws. The calculated value of K_v must be limited between 0 and 6 because the electronic motor control unit cannot handle a velocity feedback factor larger than 6. This also constrains the magnitude of the adaptation gains γ_i and γ_p. Fortunately, this has hardly any effect on the controller behaviour, as is the case for variations in γ_i and γ_p. In the actual controller, $\gamma_i = 0.4$ and $\gamma_p = 2.0$. During saturation (when the current limiters are activated), the integral part of the adaptation is stopped to avoid wind up in the adaptation integrators. During time-optimal control this situation occurs often, and so the proportional factor has the largest effect on the system response. This proportional term is mainly included to make the adaptive system react quickly to changes in the load inertia: the absence of memory in this term prevents memory of out-of-date values of the controller setting based on a previous value of J_l.

6.4 The reference model

The reference model in the adaptive controller must represent the desired time-optimal response of ω_p. This response equals that of an ideal motor model to which bang–bang control is applied, as depicted in figure 6.5. The time instant t_1, the

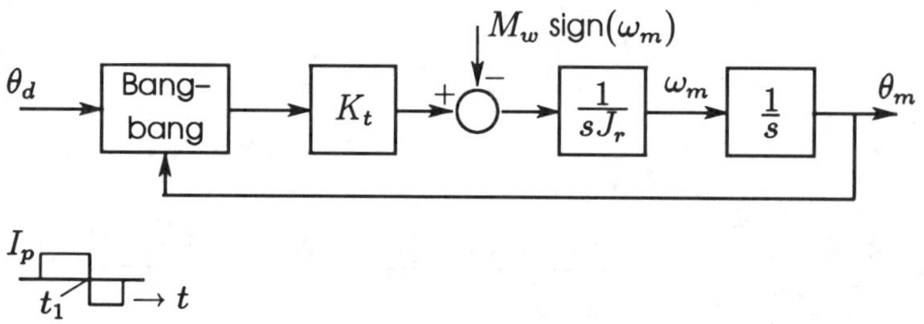

Figure 6.5 The reference model, consisting of a bang–bang controlled simplified motor model

input signal of the reference model should switch from a maximum positive to a negative value, depends on the Coulomb and viscous friction. t_1 can be calculated by compensating the maximum current for the Coulomb friction, which yields an effective positive current I_+ and a negative current I_-. Using I_+ and I_- as current limits instead of $\pm I_{\mathbf{max}}$ replaces the Coulomb friction in the reference model, and hence makes it possible to calculate the switching moment t_1. Note that the values of I_+ and I_- depend on the sign of ω_m, which is known.

Neglecting the viscous friction f, two equations must hold at the end of the movement (when $t = T$). The first equation states that at the end of the movement the velocity must be zero:

$$(K_t/J_r)\left\{\int_0^{t_1} I_+ \,\mathrm{d}t + \int_{t_1}^{T} I_- \,\mathrm{d}t\right\} = 0 \tag{6.1}$$

The second equation states that at the end of the movement the desired position must be reached:

$$(K_t/J_r)\left\{\int_0^{t_1}\int_0^{t_1} I_+ \text{ dt dt} + \int_{t_1}^{T}\int_{t_1}^{T} I_- \text{ dt dt}\right\} = \theta_d - \theta_0 \quad (6.2)$$

Assuming for simplicity that the starting angle $\theta_0 = 0$, evaluating equations (6.1) and (6.2) shows that the input signal must be switched when:

$$\theta = \theta_d \frac{-I_-}{I_+ - I_-}$$

Note that if the Coulomb friction is neglected, I_- equals $-I_+$, and so the switching point equals $\theta = \theta_d/2$ as expected. In the reference model a nominal value of the Coulomb friction is used and $I_+ = 0.883$ A; $I_- = -1.117$ A if $I_{\max} = 1$ A (see section 6.5), assuming that ω_m is positive during the movement. If ω_m were negative (when $\theta_d < \theta_p(0)$), I_+ would be 1.117 A and I_- would be -0.883 A.

The maximum acceleration in the reference model (figure 6.5), determined by J_r, should be equal to the maximum acceleration that can actually be obtained by the motor, while the current limiters make it useless to try to impose a larger acceleration. Unfortunately, the total inertia J_t, determining the maximum possible motor acceleration, is unknown. For a really time-optimal reference model J_r should be equal to J_t. To make the reference model yield a response as close as possible to a time-optimal one, the total inertia J_t is estimated by means of least-squares parameter estimation. The estimated value $\hat{J}_t$ of J_t can then be used in the reference model by copying the value of $\hat{J}_t$ into J_r.

For the estimation of J_t, we consider a parameter $\eta = 1/J_t$ which represents the transfer from the motor torque M_m (which equals $K_t \cdot I_a - M_w \operatorname{sign}(\omega_p)$ and is therefore available) to the angular acceleration α_p. This acceleration is not directly available, but is estimated by applying a difference operator $(1-z^{-1})/T_s$ to the velocity ω_p, with T_s the sampling period. This results in an estimate $\hat{\alpha}_p$ of α_p:

$$\hat{\alpha}_p = \frac{1 - z^{-1}}{T_s}\,\omega_p$$

The following least-squares algorithm is used to generate the estimate $\hat{\eta}$ of η:

$$
\begin{aligned}
\epsilon(k) &= \hat{\alpha}_p - \frac{M_m}{\hat{J}_t} \\
&= \frac{\omega_p(k) - \omega_p(k-1)}{T_s} - \hat{\eta}(k) M_m(k) \\
\hat{\eta}(k+1) &= \hat{\eta}(k) + p(k) M_m(k) \epsilon(k) \\
p(k+1) &= \frac{1}{\lambda}\left\{ p(k) - \frac{p^2(k) M_m^2(k)}{\lambda + p(k) M_m^2(k)} \right\} \qquad (6.3)
\end{aligned}
$$

The estimate $\hat{J}_t(k)$ of J_t equals $1/\hat{\eta}(k)$. In equation (6.3), λ is the forgetting factor and has a value of 0.95 during the experiments. Although the estimated acceleration $\hat{\alpha}_p$ appears to be a noisy signal, no filtering is needed because the least-squares algorithm by itself has sufficient noise-filtering properties. To avoid drift in the estimated parameter the identification procedure is switched off if the motor current I_a drops below a prespecified value. This is necessary because the excitation of the least-squares algorithm is insufficient if I_a is small.

Figure 6.6 shows an estimation result. The starting value of the estimated total inertia, $\hat{J}_t$, is set to 0.1 Nms^2/rad, while J_t actually has a value of 0.315 Nms^2/rad. Figure 6.6 shows the estimation of J_t in the top plot. The responses denoted by (a) represent the model position and velocity if no adjustment of J_t is performed, and J_t is fixed at its initial value of 0.1 (and used in the reference model). Figure 6.6 (b) shows the reference model responses which are directly dependent on $\hat{J}_t$, and (c) shows the reference model responses if J_t was known exactly. The remaining difference between the reference model response (b) and the real time-optimal response (c) is caused by the initial incorrectness of the estimate, which can be overcome by resetting the reference model to the actual velocity of the motor axis at a moment at which the estimate has reached a certain degree of accuracy. However, this procedure appears not to improve system behaviour and is not used in the actual system.

The complete control scheme incorporates two-sided adaptation. A least-squares algorithm estimates the load inertia and adjusts the reference model to the motor capabilities. Because the motor response is made to approach the reference model response, an accurate estimation of the load inertia is essential. The reference model then represents a time-optimal response of the motor, and the adaptive laws in the model reference adaptive control scheme imposes this behaviour on the motor by adjusting K_v. In the following section, some results obtained using this controller are discussed.

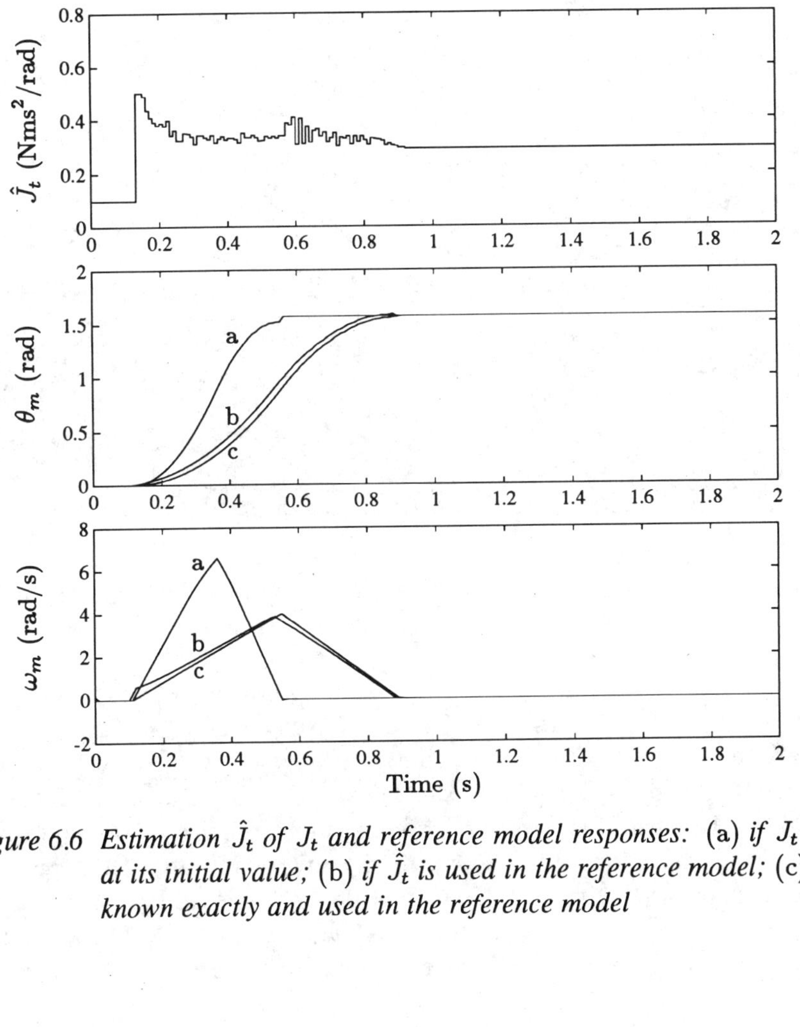

Figure 6.6 *Estimation $\hat{J}_t$ of J_t and reference model responses:* (a) *if J_t is fixed at its initial value;* (b) *if $\hat{J}_t$ is used in the reference model;* (c) *if J_t is known exactly and used in the reference model*

6.5 Experimental results

The described controller was tested on a type QT6205C motor from Inland, which is shown in figure 6.7. The following motor parameters apply to this motor (see

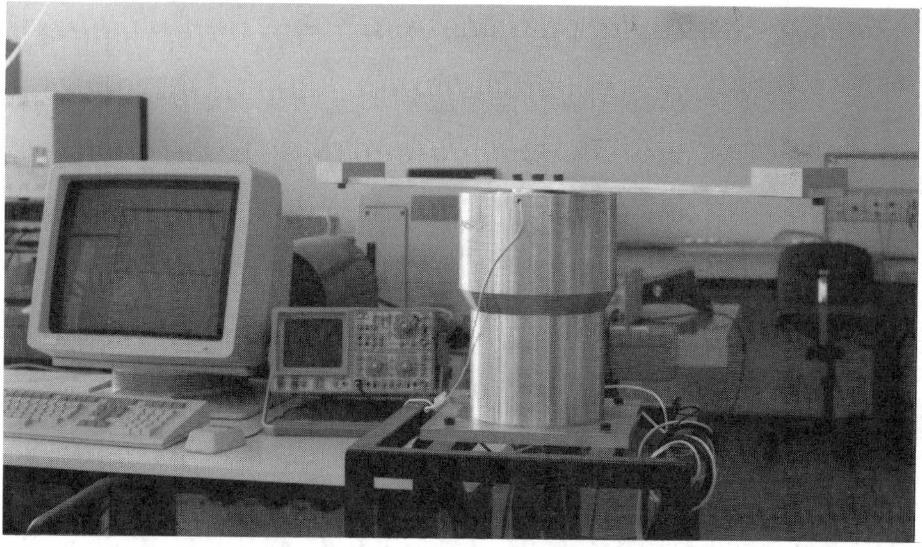

Figure 6.7 The experimental direct-drive motor with the metal arm fixed to the motor axis. The metal weights can be moved to change the load inertia

figure 6.1): $R_m = 6.67\ \Omega$, $L_m = 1.4 \times 10^{-1}$ H, $K_t = 3.26$ Nm/A, $K_b = 3.26$ Vs/rad, $J_m = 0.022$ Nms²/rad, $f = 0.013$ Nms/rad, $M_w = 0.38$ Nm, and P_{max} (max. power) is 627 W. The algorithms are implemented in **FORTRAN** on a VAXstation II minicomputer and the complete system is controlled by the real-time shell MUSIC. The hardware used for communication with the motor control unit does not allow sample times smaller than 10 ms, which is too large considering the time constants of the motor. For this reason, the peak current has been lowered to 1 A in order to reduce the motor speed. An aluminium rail has been fixed to the motor axis, to which metal weights can be bolted in different locations. This allows J_t to be varied from 0.117 to 0.460 Nms²/rad.

The proportional factor in the fixed controller is set to $K_p = 20$. To compare the PID controller and the MRAC scheme, K_i and K_v in the PID controller have been optimized for a step input of $\pi/4$ rad and a load inertia of 0.21 Nms²/rad. For this optimization the following criterion has been minimized:

$$C = \int_{t=0}^{T} \left[(\theta_p - \theta_t)^2 + (\omega_p - \omega_t)^2 \right] \, dt$$

In this formula, θ_t and ω_t represent the outputs of a real time-optimal motor model, which can be calculated because the optimization is performed for a known value of J_t. The optimization yields $K_v = 2.4$ and $K_i = 20$. In order to test the algorithm, step inputs of $\pi/8$, $\pi/4$ and $\pi/2$ rad, respectively, are applied, in combination with load inertias of 0.117, 0.210 and 0.460 Nms^2/rad. For each combination, the above-stated criterion is evaluated for both the PID and the MRAC controller. The results are listed in table 6.1.

θ_d J_t	$\pi/8$	$\pi/4$	$\pi/2$	
0.117	0.087	0.096	0.073	PID
Nms^2/rad	0.022	0.017	0.055	MRAC
0.210	0.025	**0.007**	0.253	PID
Nms^2/rad	0.004	**0.003**	0.042	MRAC
0.460	0.023	0.169	1.211	PID
Nms^2/rad	0.012	0.014	0.014	MRAC

Table 6.1 Criterion values without adaptation (PID) and with adaptation (MRAC). The boldface numbers denote the values for the combination of step size and load inertia for which the PID controller is optimized

If $\theta_d = \pi/4$ rad and $J_t = 0.21$ Nms^2/rad, the criterion value for both the PID and MRAC controller is small, due to the fact that the PID controller is optimized for this combination. The responses in this case are similar. However, if the load inertia is altered, the performance of the PID controller deteriorates quickly, whereas the performance of the MRAC controller remains good. A change in the magnitude of the step input has the same effect.

Obviously, the MRAC controller can improve the motor behaviour considerably in the scope of this criterion. As shown in figures 6.8 and 6.9, where $\theta_d = \pi/8$ rad and $J_t = 0.117$ Nms^2/rad, the PID controlled motor is slower than the MRAC controlled system. In these diagrams, a *real* time-optimal response, denoted by θ_t and ω_t (which is independent of estimation errors in J_t), is shown for comparison. Under extreme conditions the difference is even more apparent; figures 6.10 and 6.11 show the motor behaviour if $\theta_d = \pi/2$ and $J_t = 0.460$ Nms^2/rad. The MRAC scheme now prevents the motor response from large overshoot.

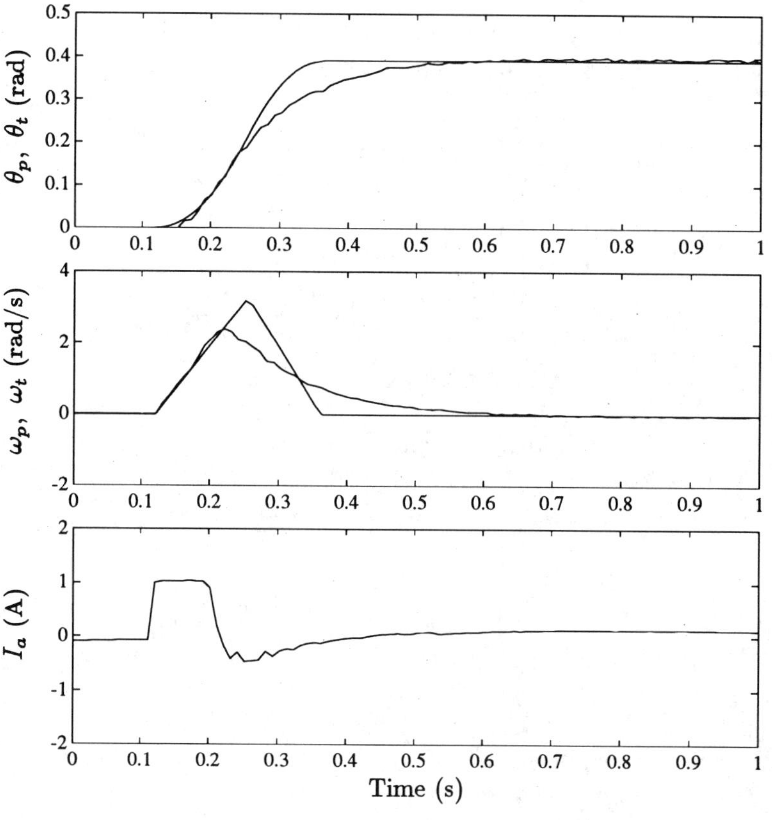

Figure 6.8 Motor response obtained with $\theta_d = \pi/8$ rad and $J_l = 0.117$ Nms^2/rad using PID, and time-optimal desired behaviour

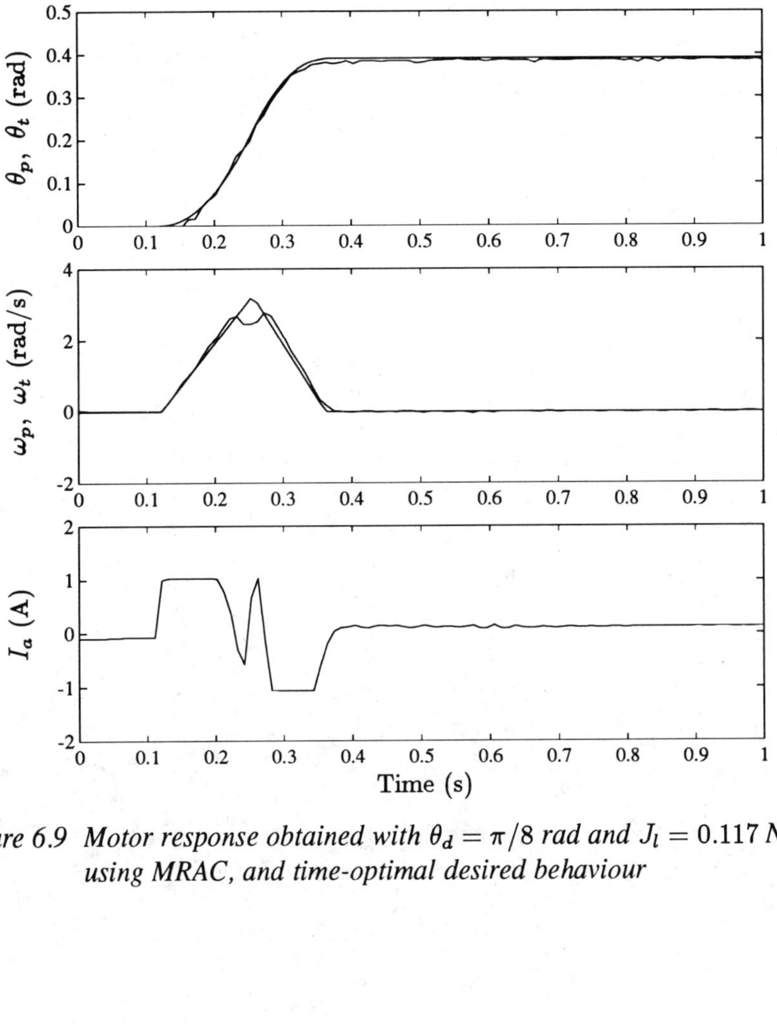

Figure 6.9 Motor response obtained with $\theta_d = \pi/8$ rad and $J_l = 0.117$ Nms^2/rad using MRAC, and time-optimal desired behaviour

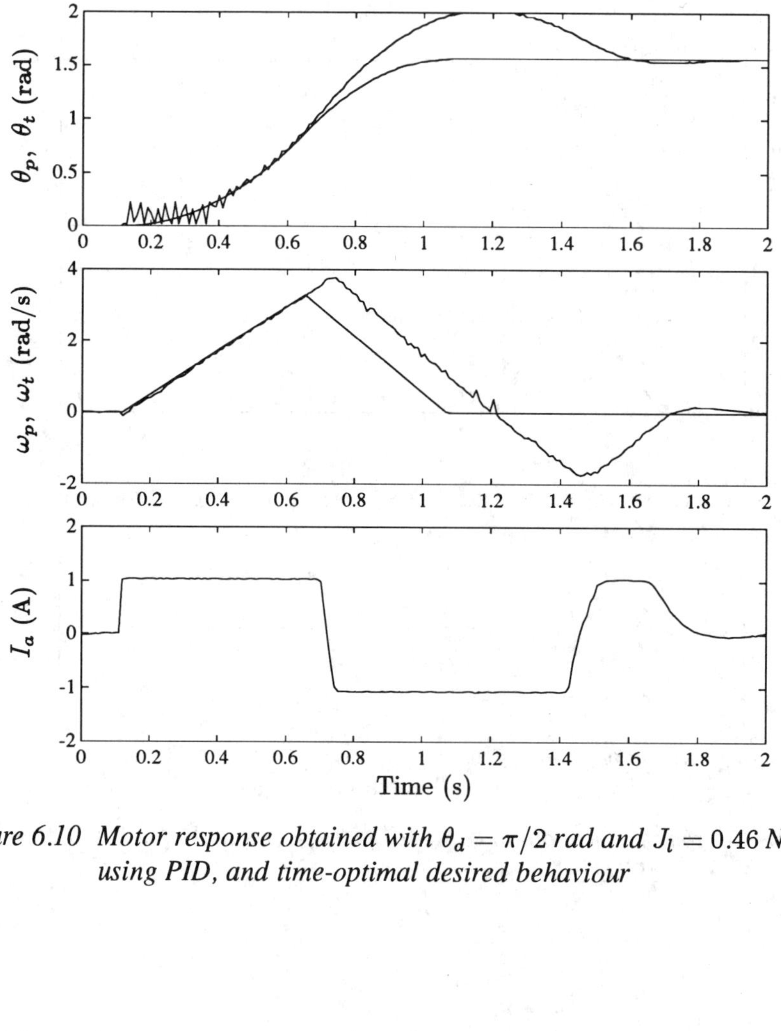

Figure 6.10 Motor response obtained with $\theta_d = \pi/2$ rad and $J_l = 0.46$ Nms²/rad using PID, and time-optimal desired behaviour

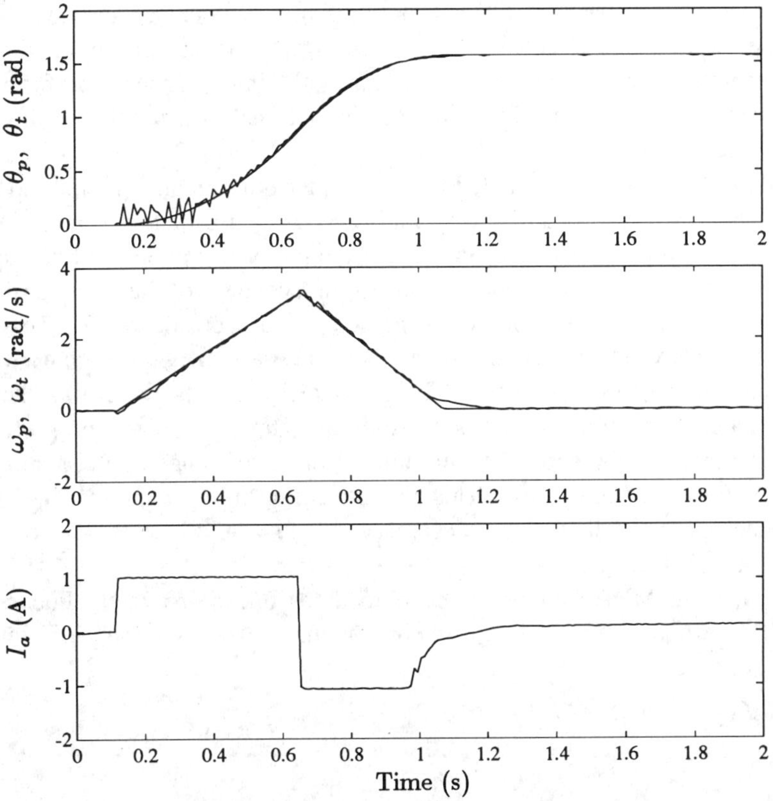

Figure 6.11 *Motor response obtained with $\theta_d = \pi/2$ rad and $J_l = 0.46$ Nms^2/rad using MRAC, and time-optimal desired behaviour*

6.6 Summary

Adaptive model adjustment is useful when the control specifications cannot be met using a fixed reference model, even if the process can be made to follow the model exactly. This problem occurs mainly when limits on the process input are very apparent, and hence the maximum rate of change of the output depends on (some of) the process parameters. A fixed reference model would have to be chosen such that the process can always follow the model exactly, and hence would have to be as slow as the slowest possible process response. On-line estimation of the parameters that determine the maximum obtainable process performance, and use of these parameters in the reference model, makes it possible to improve the overall performance.

While in the example presented in this chapter only one parameter had to be estimated, this estimation can be very rapid. Making use of the estimate of the inertia in the reference model guarantees that the motor behaviour will be approximately time optimal if a step input is applied, independent of the load inertia and the set-point magnitude. The proposed method can be considered as 'two-sided' adaptation. The explicit estimation of the load inertia makes the scheme resemble an indirect adaptation scheme. The estimate is used to determine the best reference model for the actual motor capabilities. Through a direct proportional–integral adaptive law the approximately time-optimal reference model response is imposed on to the actual process. This way, no compromise regarding the control goal (a time-optimal behaviour) is necessary, which would be the case if a fixed reference model were used.

Note that all MRAC theory is developed for linear or exactly linearizable processes. The need for model adjustment was introduced as a result of violation of this requirement.

Problem

In section 6.3, it was mentioned that the angle θ_p (or, rather, the error between θ_p and θ_m) need not be included in the adaptive law because only K_v is adjusted. Now, consider the somewhat simplified system depicted in figure 6.12.

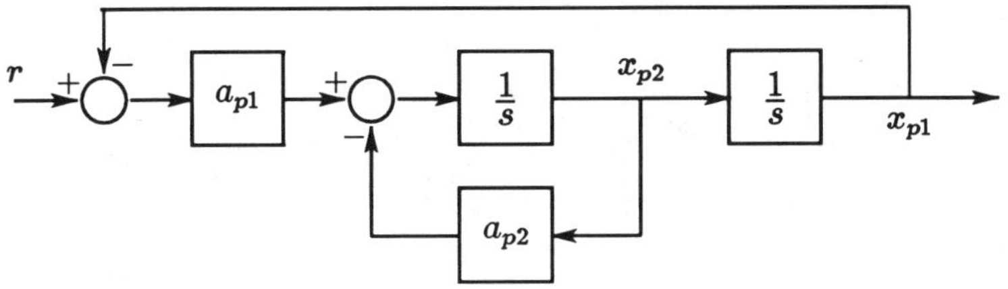

Figure 6.12 Simplified controlled motor transfer function

Assume that the reference model is given by:

$$\dot{\boldsymbol{x}}_m = \boldsymbol{A}_m \boldsymbol{x}_m + \boldsymbol{b}_m r$$

with:

$$\boldsymbol{A}_m = \begin{pmatrix} 0 & 1 \\ -a_{m1} & -a_{m2} \end{pmatrix}$$

and:

$$\boldsymbol{b}_m = \begin{pmatrix} 0 \\ a_{m1} \end{pmatrix}$$

Hence, the reference model has a DC gain of 1 as has the closed-loop process. Further, assume that a_{p1} is equal to a_{m1}, and hence needs no adjustment. Show that in adjusting a_{p2}, an error signal $e_2 = x_{p2} - x_{m2}$ suffices and use of $e_1 = x_{p1} - x_{m1}$ is not required.

7

Direct Model Adjustment: an Application

Section 2.2 presented a MRAC scheme for multi-input, multi-output processes which was based on the availability of the state vector of all subprocesses. This made it possible to satisfy the perfect model-matching condition with a straightforward extension of state feedback. If the outputs of the subprocesses only are available, such a procedure is not possible. For the general MIMO case formal solutions are not yet readily available.

This chapter presents a method for the control of coupled systems of which only the outputs are available, serving as an illustration of alternative uses of model adjustment. The method primarily involves consideration of the interaction as a disturbance for each of the subprocesses. Although the method to be presented is mainly heuristic, examples illustrate its usefulness.

7.1 Problem statement

This chapter considers the situation depicted in figure 7.1. The process $W_p(s)$ is disturbed by an external disturbance which is correlated with the reference input r. The disturbance may enter the process at any point; in figure 7.1 the effect of the disturbance on the process output y_p is denoted $\nu(r)$. The case depicted in figure 7.1 occurs in interacting systems; $\nu(r)$ is then generated by a second (possibly adaptive) system which is also driven by r (Butler and Soeterboek, 1991).

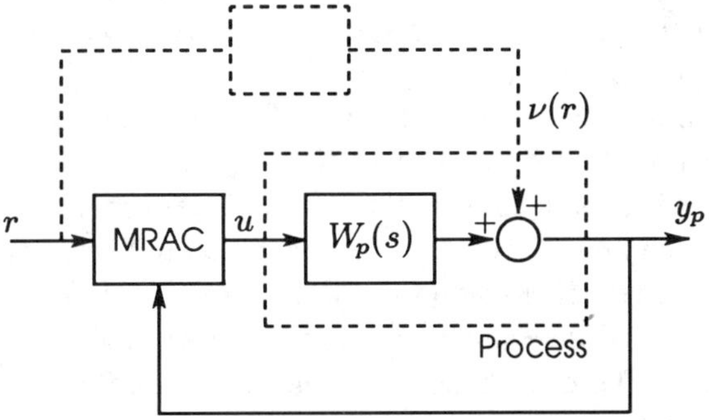

Figure 7.1 General problem setting: the process output disturbance ν is correlated with the reference r

Assuming a controller structure of the augmented error method, y_p can be written as:

$$y_p = W_p\left[\theta^T\left(\omega + \omega^{\nu}\right)\right] + \nu(r) = W_p\left(\theta^T\omega\right) + \nu'(r) \tag{7.1}$$

In equation (7.1), ω^{ν} denotes the part in the signal vector that is caused totally by penetration of the disturbance ν, and hence the actual signal vector consists of the addition of the original ω and ω^{ν} (as in chapter 3). The output error becomes:

$$\begin{aligned} e_1 &= y_p - y_m = W_m\left(\phi^T\omega\right) + \nu'(r) \\ &= e_1' + \nu'(r) \end{aligned}$$

where e_1' is the original output error if no disturbance is present. Assuming, for a moment, that no error augmentation or compensation is required, in the adaptive scheme the output error is multiplied by the signal vector:

$$\begin{aligned}\dot{\boldsymbol{\theta}} &= -\boldsymbol{\Gamma}\,(e_1' + \nu')\,(\boldsymbol{\omega} + \boldsymbol{\omega}^{\nu}) \\ &= -\boldsymbol{\Gamma}\,(e_1'\boldsymbol{\omega} + e_1'\boldsymbol{\omega}^{\nu} + \nu'\boldsymbol{\omega} + \nu'\boldsymbol{\omega}^{\nu}) \qquad (7.2)\end{aligned}$$

The first term in equation (7.2) is the original term, as desired. If the disturbance is not correlated with the undisturbed signal vector (e.g. when ν is stochastic and has zero mean) the second and third terms are zero. Only the last term introduces an offset at the integrator inputs if ν' is correlated with $\boldsymbol{\omega}^{\nu}$, which is usually true. In the case at hand, both ν' and $\boldsymbol{\omega}$ depend directly on the reference signal r, and so they are mutually correlated. The same is true for e_1' and $\boldsymbol{\omega}^{\nu}$, and so all cross terms contribute to the integrator offsets.

If there is knowledge of the external disturbance caused by the interaction, one may wonder how to use this knowledge. Inspecting equation (7.2), it can be seen that $(\boldsymbol{\omega}+\boldsymbol{\omega}^{\nu})$ is generated by the auxiliary signal generators from the process output y_p and the control u. No direct influence on this term can be exerted. To decrease the cross terms in equation (7.2), the only way is, therefore, to decrease the effect of ν on e_1. This can be achieved by direct model adjustment, as introduced in chapter 5. The disturbance is included in the reference model response, in such a way that the reference model transfer function still yields the desired behaviour, but has a similar disturbance effect as the process output. This way, the influence that ν has on the output error can be diminished.

The system generating $\nu(r)$ may, for example, consist of a MRAC-controlled process. In such a case, one is actually dealing with a multi-input, multi-output system in which the interaction is considered to be an external disturbance for each of the processes separately. Such a simplified view on adaptive control of MIMO processes differs completely from that discussed in section 2.2, where the design was based on knowledge of the complete system. However, the adaptive control problem for MIMO systems of which only the outputs are known is not yet completely solved (Narendra and Annaswamy, 1989; Sastry and Bodson, 1989), and in addition can serve as an illustration of the large potential of model adjustment.

7.2 An example

One-sided interaction and two-sided (or mutual) interaction have fundamentally different effects on the behaviour of an adaptive system. One-sided interaction can be regarded a form of external disturbance, and although the disturbance is a little more complicated than those considered in chapter 3, its effect is comparable. A performance decrease is to be expected, but severe stability problems are not likely. Mutual interaction, however, introduces extra time-varying unmodelled dynamics in the adaptation loops, and is therefore more likely to cause serious problems. This section presents an example of one-sided interaction. A mutual interaction problem is described in section 7.3, where the method is applied to a practical problem.

In this example, two processes are both controlled by an MRAC system. The output of process 1 disturbs the output of process 2, hence introducing one-sided interaction. The process and model transfer functions are chosen as follows:

$$\text{system 1: } W_{p1} = \frac{1}{s+1}; \qquad W_{m1} = \frac{3}{s+3}$$

$$\text{system 2: } W_{p2} = \frac{2}{s(s+8)}; \qquad W_{m2} = \frac{16}{s^2+8s+16}$$

The process output y_{p2} is disturbed by $\nu = \delta_p y_{p1}$, as shown in figure 7.2. If δ_p is zero, and hence there is no interaction, the result given in figure 7.3 is obtained for y_{p1} and y_{p2} by using the augmented error method separately for both

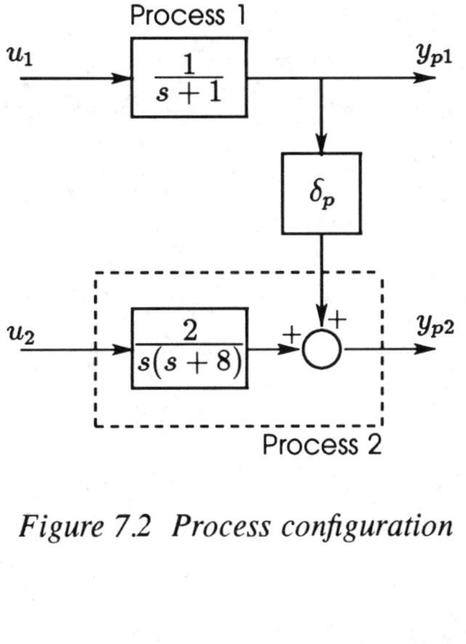

Figure 7.2 Process configuration

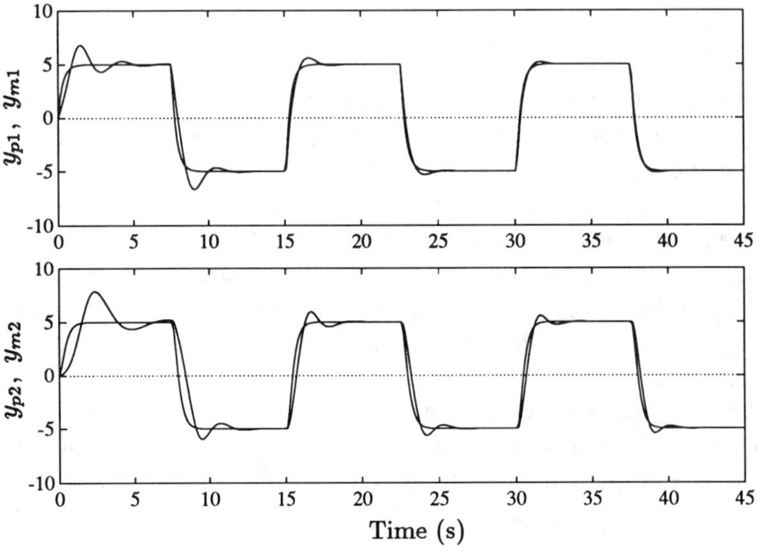

Figure 7.3 Response of y_{p1} and y_{p2}, obtained if no interaction is present

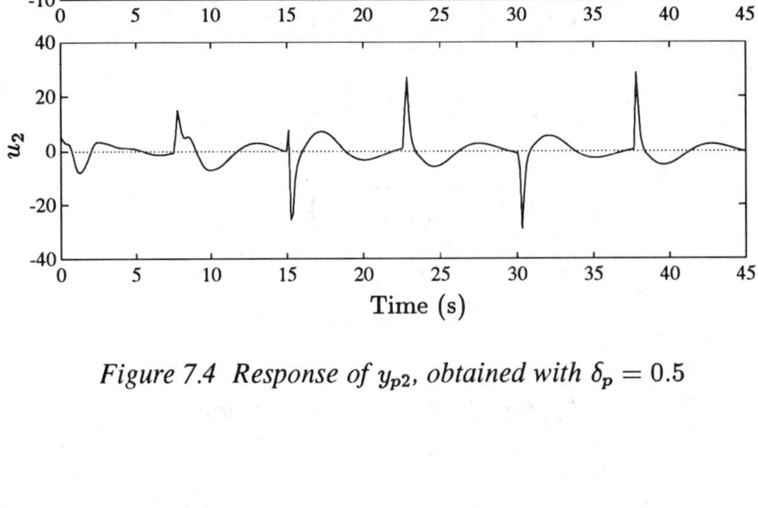

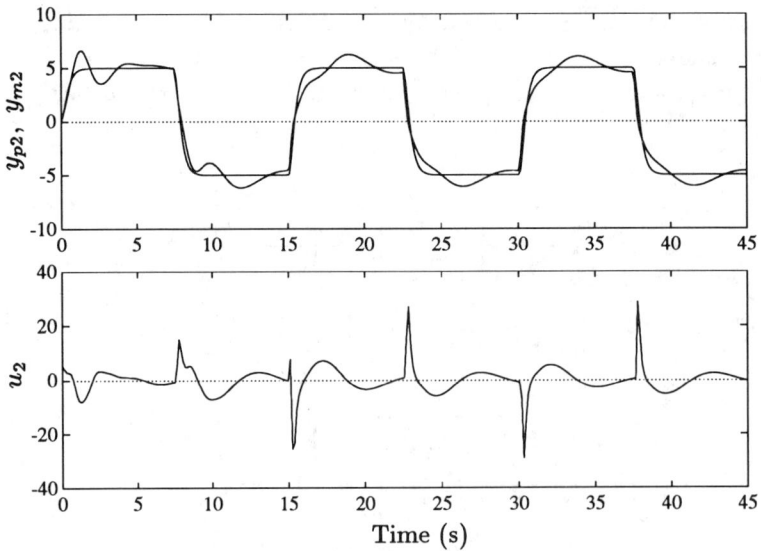

Figure 7.4 Response of y_{p2}, obtained with $\delta_p = 0.5$

processes. If $\delta_p = 0.5$, the response shown in figure 7.4 appears for y_{p2}. Because the interaction is one sided only, y_{p1} is the same as in figure 7.3. It can be seen that the disturbance during the transients prohibits the adaptation from converging properly, and the parameter values have to be adjusted each time the reference signal has changed.

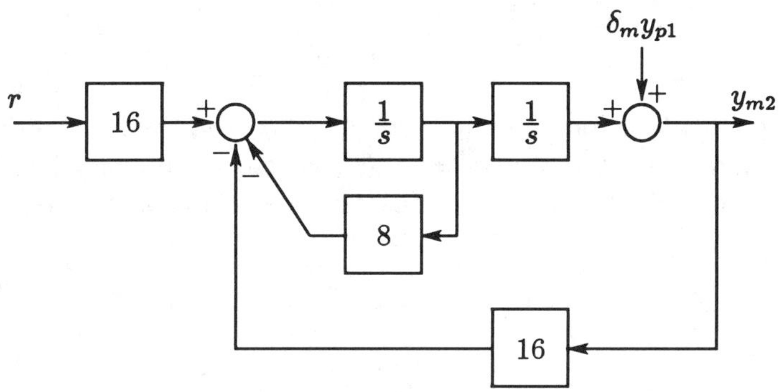

Figure 7.5 Disturbance injection into reference model

To apply the reference model adjustment, a signal $\delta_m y_{p1}$ is input in the reference model as well. Because the reference model should still give a desired response, the point at which the disturbance acts on the model must be chosen with care. In this case the disturbance is injected as shown in figure 7.5. The transfer function from $\delta_m y_{p1}$ to y_{m2} is:

$$\frac{y_{m2}}{\delta_m y_{p1}} = \frac{s^2}{s^2 + 8s + 16}$$

The poles of this transfer function are determined by the reference model, and hence are properly damped. In addition, for steady-state disturbances no steady-state error on the reference model output occurs.

Analyzing the effect on the error equation in a general way is difficult because the effect of ν on y_{p2} depends on the current parameter setting $\boldsymbol{\theta}_2$. Therefore, it is assumed for a moment that $\boldsymbol{\theta}_2 = \boldsymbol{\theta}_2^*$: $k_{20}^* = 8$, $c_{21}^* = 0$, $d_{20}^* = -8$ and $d_{21}^* = 0$. Then:

$$\frac{y_{p2}}{\nu} = \frac{s(s+8)}{s^2 + 8s + 16} \tag{7.3}$$

Because:

$$\frac{y_{m2}}{\nu} = \frac{s^2}{s^2 + 8s + 16}\left(\frac{\delta_m}{\delta_p}\right) \tag{7.4}$$

the effect of ν on the output error becomes:

$$\nu' = \left(\frac{1}{\delta_p}\right)\frac{s(s+8)\delta_p - \delta_m s^2}{s^2 + 8s + 16}$$

The norm of this function is minimal for $\delta_m = \delta_p$. Figure 7.6 shows the effect of ν on the error equation for $\delta_m = 0$ ((a), no model correction) and $\delta_m = \delta_p$ ((b), optimal model correction) as a function of the frequency. It can be seen

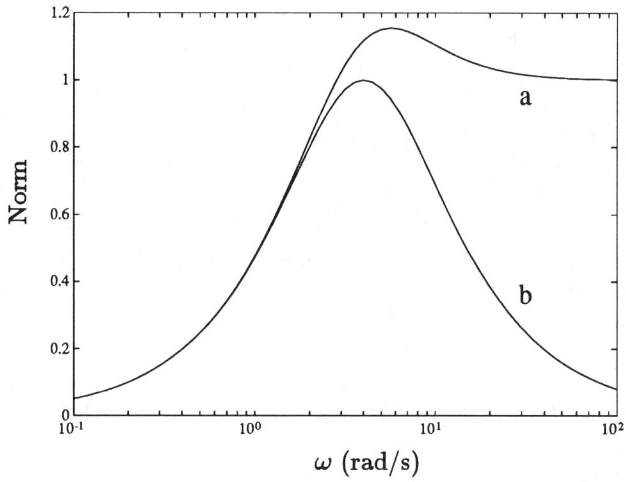

Figure 7.6 Effect of ν on output error for: (a) $\delta_m = 0$, *and* (b) $\delta_m = \delta_p$

that, especially in the high-frequency range, the effect of the model adjustment is considerable. This means that, especially during the transients, which contain high frequencies, the model adjustment is expected to be useful, which is the purpose of the adjustment. In addition, the improvement will mainly occur if the process inducing the disturbance has small time constants, in which case the effect of the disturbance on y_{p2} is greatest (figure 7.6 (a)). Figure 7.7 shows a simulation result for $\delta_m = \delta_p = 0.5$. It can be seen that the response is improved considerably compared with that shown in figure 7.4. However, the error signal does not tend to zero exactly, because ν' can never be made zero, even with implementation of the model adjustment. This inability to satisfy the perfect model-matching condition

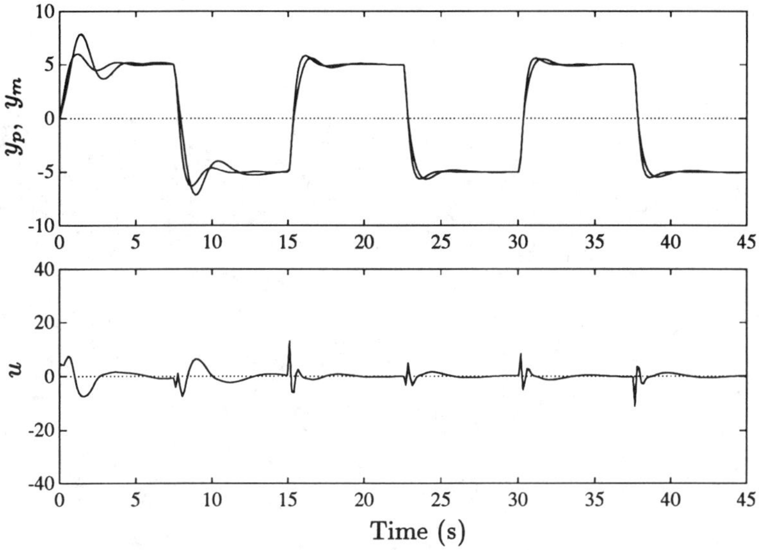

Figure 7.7 Result obtained by optimal model adjustment ($\delta_m = 0.5$)

follows from figure 7.6, in which the norm is nonzero for all $\omega > 0$. An additional effect is the reduced control effort, as was the case in the decomposition method discussed in chapter 5.

If the interaction between two processes is mutual, the problem is more difficult because, for each of the processes, the interaction causes unmodelled dynamics due to the extra created loop. However, a procedure similar to that presented for one-sided interaction can be applied: the disturbance occurring is injected into each of the reference models. Of course, in both the one-sided and the mutual interaction case, this procedure can only be applied if the disturbance is known or can be derived indirectly from measurable signals.

7.3 Application to a helicopter propeller setup

The method described above has been tested on an experimental propeller setup, a photograph of which is shown in figure 7.8. A schematic diagram is shown in figure 7.9. A metal arm is connected to its base by two ball bearings, such that it can move freely both in the horizontal and vertical directions. At both ends, propellers are fitted at distances l_1 and l_2 from the base. Normally, the propellers are positioned perpendicularly to each other such that movement in the

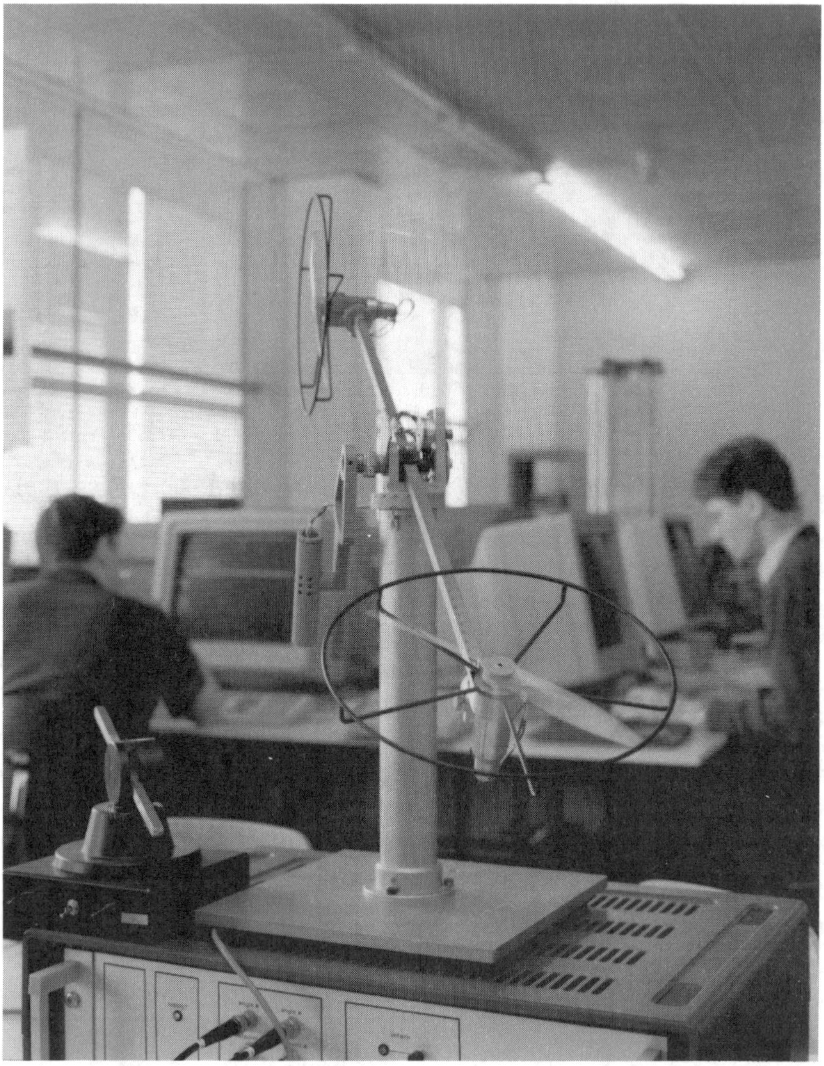

Figure 7.8 The propeller setup, with both propellers in their standard position

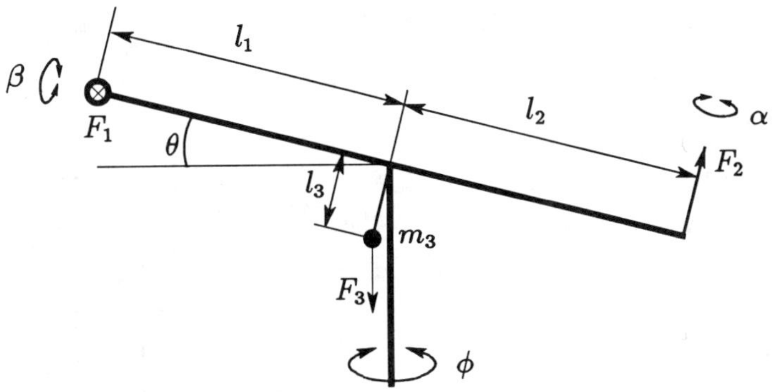

Figure 7.9 Schematic view of the propeller setup

vertical plane (angle θ) and in the horizontal plane (angle ϕ) is effected by only one propeller. In this case, the angles α and β indicating the deviation of the θ-propeller and ϕ-propeller, respectively, from the standard position, are both zero. At the connection of the arm with the base, a weight with mass m_3 is connected to the arm at a distance l_3. In the base, potentiometers are fitted which measure the angles θ and ϕ, and at the motors tacho generators provide the rotation speed of the propellers. The control problem is to make the arm move quickly and accurately to a specified attitude, given in the form of set points for θ and ϕ (note that, in this section, ω denotes angular velocity, and θ and ϕ denote angles).

7.3.1 Mathematical model of the propeller setup

A block diagram of the propeller transfer function is shown in figure 7.10. The control voltages U_1 and U_2 are input to the motors which drive the propellers. The motor transfer functions, neglecting the electrical time constants, are:

$$\frac{\omega}{U} = \frac{25}{0.52s + 1} \tag{7.5}$$

in which ω is the angular velocity. The maximum allowable control voltage is ± 10 V. The angular velocity of the motors results in a force F on the arm perpendicular to the propellers, which in turn results in a movement of the arm. The relation between the propeller velocity and the resulting force is nonlinear, and can be approximated by: $F(\omega) = \text{sign}(\omega) \times 3.8 \times 10^{-6}\omega^2$. Note that the form of the propeller is not completely symmetric so that the behaviour in one direction

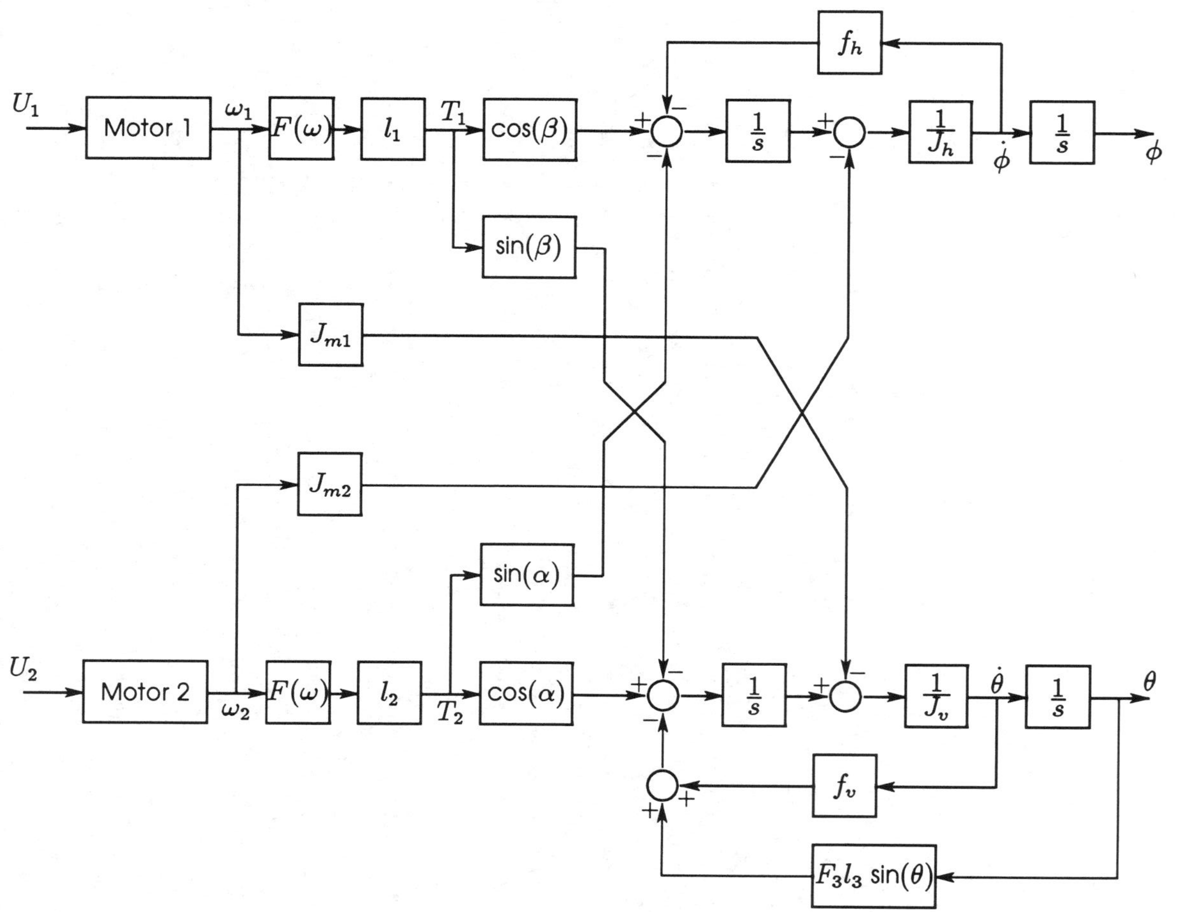

Figure 7.10 Block diagram of the propeller transfer function

is slightly different from that in the other direction; the constant factor 3.8×10^{-6} is used instead of 3.2×10^{-6} and 4.3×10^{-6} for the two directions.

Rotation of a propeller produces an angular momentum which, according to the law of conservation of angular momentum, must be compensated by the remaining body of the propeller stand. This causes interaction between the two transfer functions, represented by the moments of inertia of the motors and propellers, J_{m1} and J_{m2}. This interaction directly influences the velocities $\dot{\theta}$ and $\dot{\phi}$. Note that if $\alpha \neq 0$ or $\beta \neq 0$, J_{m1} also affects $\dot{\phi}$ and J_{m2} also affects $\dot{\theta}$ (through factors $\sin(\beta)$ and $\sin(\alpha)$ respectively), which is not depicted in figure 7.10. In the following, interaction caused by J_{m1} and J_{m2} is not considered because its influence is relatively small.

Multiplication of the forces F_1 and F_2 by the arm lengths l_1 and l_2 produces the torques acting on the arm. If α and β are not zero and hence the propellers are twisted, interaction between the two transfer functions arises, represented by the sine and cosine terms in figure 7.10. The resulting torques are input to second-order systems which represent the mechanical transfer from the torques to the angles θ and ϕ. The transfer function for the horizontal movement ϕ has a pure integrating term while the transfer function for θ has not, due to the mass m_3. The parameters J_v and J_h are the moments of inertia of the arm and propellers in vertical and horizontal directions respectively. These values are slightly different due to the mass m_3. f_v and f_h represent the viscous friction for the two directions. The following values for the various parameters apply:

$$
\begin{aligned}
l_1 &= 0.316 \text{ m} \\
l_2 &= 0.316 \text{ m} \\
J_v &= 18 \times 10^{-3} \text{ Nms}^2/\text{rad} \\
J_h &= 15.2 \times 10^{-3} \text{ Nms}^2/\text{rad} \\
f_v &= 6 \times 10^{-3} \text{ Nms/rad} \\
f_h &= 7.6 \times 10^{-3} \text{ Nms/rad} \\
m_3 g l_3 &= 30 \times 10^{-3} \text{ Nm}
\end{aligned}
$$

Further, the tacho sensitivity is 8.3×10^{-3} Vs/rad and the potentiometer sensitivities are 19 V/rad and 3.6 V/rad for the angles θ and ϕ respectively. The turning ranges for θ and ϕ are $\pm 30°$ and $\pm 150°$ respectively. The model as presented in this section is only a limited representation of the real system. For example, the fact that the motor centres of inertia are not completely in line with the arm makes the model more complicated. These and other effects have been neglected.

7.3.2 The primary controller

In the absence of interaction, both transfer functions are of order 3. Using the augmented error method would require six parameters to be adjusted, which in practice is not favourable due to the presence of measurement noise and the expected lack of convergence. One way to use a lower-order controller is to consider a part of the system dynamics as unmodelled, and use the decomposition method as presented in chapter 5. However, the tacho generators for the propeller motors make it possible to implement an inner loop containing speed control for both motors. In this way the response of the motor velocity with respect to the control voltage can be made very fast, leading to a wide separation between the motor time constants and the time constants of the mechanical part. Such a wide separation allows the neglect of the motor time constants altogether, making the use of a lower-order MRAC controller possible without further modifications. As a compromise between the minimum sampling period of 10 ms, the control limits of ± 10 V and the desired high motor speed, the velocity loop for both motors was selected:

$$U_{\mathbf{motor}} = K_p (U - K_t \omega)$$

where $K_p = 10$ and $K_t = 2.1$ (including a tacho constant of 8.3×10^{-3} Vs/rad and interface constants of 2.41 for the tacho voltage and 1.2 for the control voltage). This leads to a closed-loop motor transfer of:

$$\frac{\omega}{U} = \frac{48}{0.1s + 1}$$

The motor transfer functions are now considerably faster than the mechanical part, which has the additional advantage that a simple correction for the nonlinearity in the propeller transfer function can be implemented. As stated above, the force applied by the propeller on the arm is proportional to the square of the propeller velocity. Since the transfer from the control voltage to the propeller velocity is relatively fast, this nonlinear effect can be compensated for by taking the square root of the control voltage and applying the result to the motor inputs. In other words, the following approximation is allowed:

$$F = \left[K_{DC} H_{\mathbf{motor}}(s) \sqrt{U}\right]^2 \approx K_{DC}^2 H_{\mathbf{motor}}(s) U \tag{7.6}$$

In equation (7.6), the DC gain of the motor transfer function is denoted K_{DC}, while $H_{\mathbf{motor}}$ has a DC gain of 1. The purpose of this modification in the control input is that of being able to consider the system linear for the MRAC design. Both the θ-system and the ϕ-system are then assumed to be of second order.

7.3.3 The adaptive controller and the reference model

The augmented error method has been implemented for both directions separately. The reference model plays a crucial role in this implementation. To be able to take the angles α and β into consideration at a later stage, the structure of the reference model has been made equal to that given in figure 7.10. The reference model includes the motor transfer functions with the velocity controllers, input limitation at ± 10 V, and the propeller nonlinearity with its compensation. Thus, a nonlinearity effect which has not been completely eliminated is present in the reference model output also, making the error signal insensitive to this effect. The reference model is equipped with state feedback of θ, $\dot{\theta}$ and ϕ, $\dot{\phi}$. The feedback parameter values were determined by optimization of the model in simulation, using the criterion functions:

$$C_\theta = \int_{t=0}^{T} t\,|\theta_d - \theta|\,\mathrm{d}t$$

$$C_\phi = \int_{t=0}^{T} t\,|\phi_d - \phi|\,\mathrm{d}t$$

where θ_d and ϕ_d are the set points for θ and ϕ respectively. By weighting the time in the criterion function, the overshoot in the transfer functions is almost completely eliminated. To avoid steady-state errors for different set points, the term $\sin(\theta)$ is approximated by θ in the model transfer function for θ. Because the turning range of θ is relatively small, this approximation is allowed.

By choosing the reference model structure equal to the structure of the physical process, the maximum flexibility in alteration or modification of the model is created. However, calculation of the filters L^{-1} in the augmented error method has now become impossible because the reference model contains nonlinearities and hence it is not possible to investigate the SPR property of $W_m L$. Therefore, for the calculation of the filters L^{-1}, the model transfer functions are approximated by third-order linear transfer functions, on the basis of which calculation of L^{-1} is performed.

7.3.4 Experimental results

No interaction

The MRAC algorithm as described above is first applied to the 'standard' system with $\alpha = \beta = 0$. Again, the system is operated using a VAXstation II mini-computer with the MUSIC shell. Both controllers are designed for a system of order 2, and so have four parameters to be adjusted. Use of the optimal feedback parameters in the reference model as described above causes some problems because of the neglect of J_{m1} and J_{m2}. Therefore, the model is made slightly slower by increasing the velocity feedback. The result is shown in figure 7.11. In the top window is shown the response of ϕ to a block input signal with an amplitude of 32°, and the bottom window shows the response of θ to a block input with an amplitude of 18°.

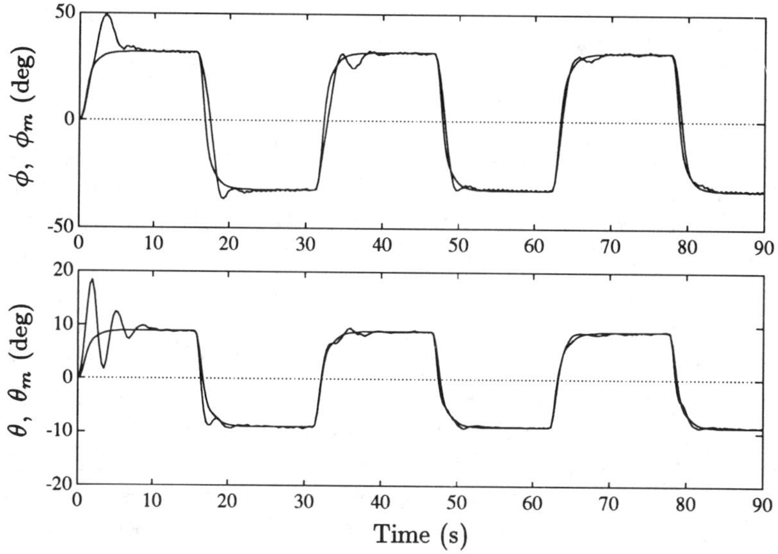

Figure 7.11 Helicopter response without interaction

One-sided interaction

In the next experiment the ϕ-propeller is turned such that rotation of this propeller not only moves the arm in the horizontal direction but also causes movement in the vertical direction. This implies that a one-sided interaction is created: movement of ϕ disturbs the transfer function for θ, but θ leaves ϕ undisturbed. The angle β is chosen as 30°, and this value for β is also incorporated in the reference model. The application of MRAC without modifications yields the result shown in figure 7.12.

As in the example given in section 7.2, the closed-loop transfer function of the disturbed process does not converge to the reference model. During each set-point change, the parameter adaptation is disturbed, and it takes some time before θ reaches its desired set point. Note that the behaviour of ϕ is influenced slightly as well, which is caused partly by the changed gain in the feedforward path of ϕ (due to the term $\cos(\beta)$), and partly by the neglected coupling factors J_{m1} and J_{m2}.

To deal with the interaction problem, the method described in section 7.2 is implemented by injecting the disturbance into the reference model also. In this case, the disturbance is formed by $\sin(\beta)T_1$, with T_1 the torque produced by the ϕ-propeller. This torque is not measured, but is calculated from the measured motor velocity ω_1 using the propeller nonlinearity $F(\omega)$. The calculated disturbance torque is injected into the reference model for θ at the position of the term $\sin(\beta)$ in figure 7.11. Note that in steady state the disturbance is zero because the transfer function for ϕ is a type 1 system and so the propeller torque is then zero. Therefore, the reference model transfer function does not have a steady-state error resulting from the disturbance injection. The result of this new approach is shown in figure 7.13. Although the response of θ still suffers from overshoot, the steady state is reached more quickly.

Two-sided interaction

In the last experiment both propellers are turned, with angles $\alpha = -30°$ and $\beta = 30°$. The interaction is now mutual, and for each of the loops the effect has the form of time-varying unmodelled dynamics. Figure 7.14 shows that the mutual interaction causes instability if no modification is used. In implementing the model adjustment as presented in section 7.2, the disturbance calculated from the tacho voltages is injected in both the θ-reference model and the ϕ-reference model. Figure 7.15 shows that the approach has succeeded: the model is not followed perfectly but the result is much better than without disturbance injection.

The disturbance injection makes the output error for both transfer functions

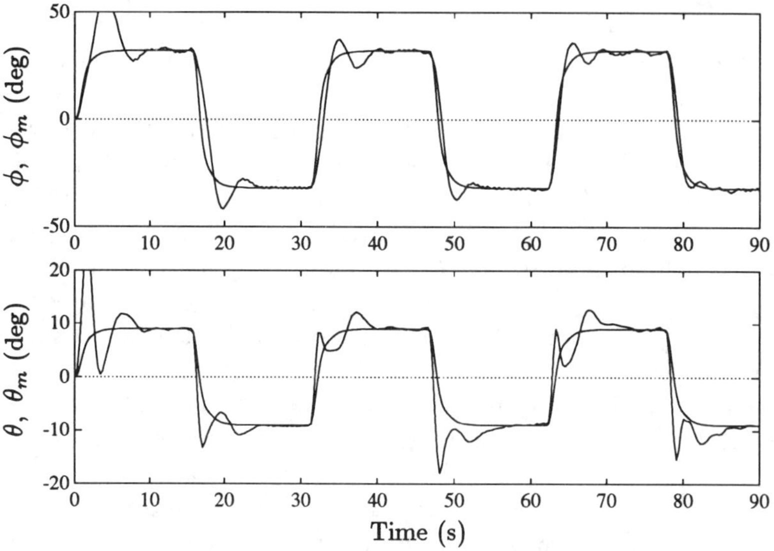

Figure 7.12 Helicopter response if $\beta = 30°$, obtained without modification

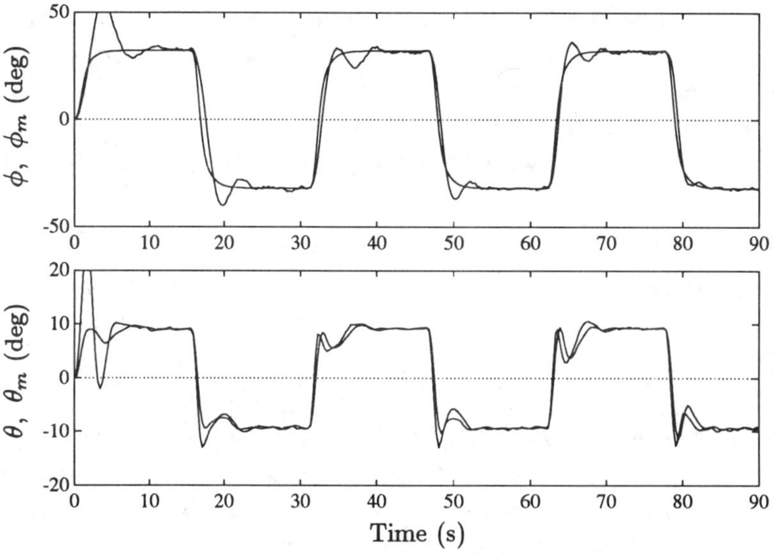

Figure 7.13 Helicopter response if $\beta = 30°$, obtained by using disturbance injection into reference model

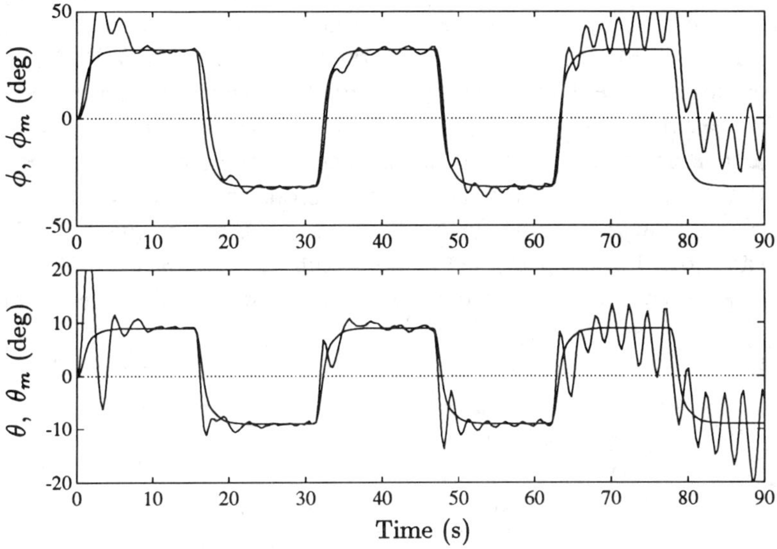

Figure 7.14 Helicopter response if $\alpha = -30°$ and $\beta = 30°$, obtained without modification

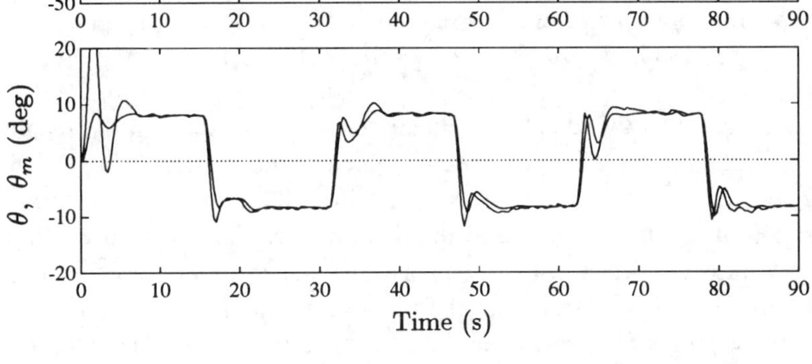

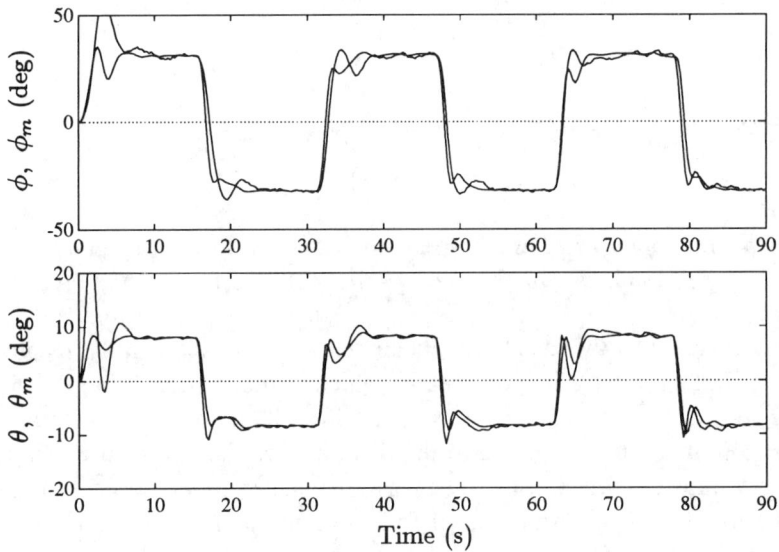

Figure 7.15 Helicopter response if $\alpha = -30°$ and $\beta = 30°$, obtained by using disturbance injection into the reference model

less sensitive to the disturbance by including the disturbance in the reference model also. Another approach is to attempt direct compensation of the effect of the disturbance in the process. This way the reference model is left unchanged but the expected disturbance, known from the reference model, is added to the control signal. For example, the torque produced by the ϕ-propeller in the reference model is a measure of the expected disturbance in the actual process. Because the path from the control signal to the effective torque for θ is almost direct (only the dynamics of the motor transfer function, which have relatively small time constants, are placed in the path), the real effect of the disturbance in the process can be compensated directly by an extra input signal at the θ-loop. In figure 7.16

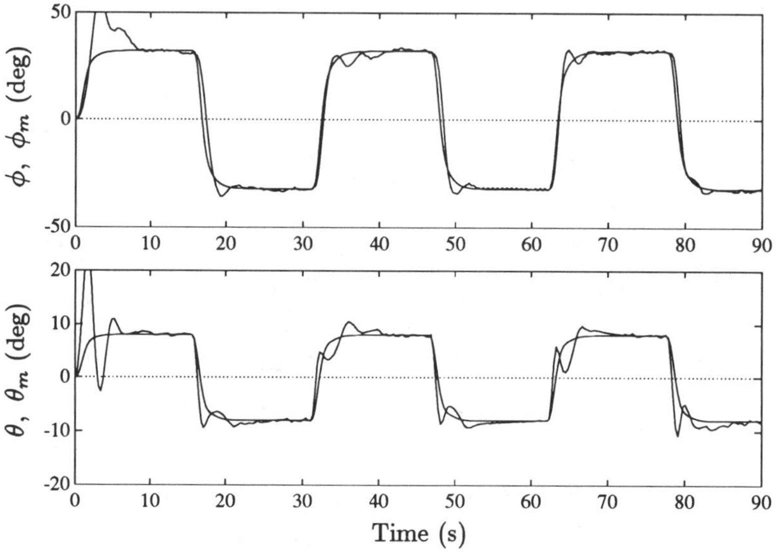

Figure 7.16 Helicopter response if $\alpha = -30°$ and $\beta = 30°$, obtained by injection of a compensation for the expected disturbance into the process

it can be observed that the instability phenomenon is avoided and the response is fairly good. In fact, the method of injecting an extra model signal into the process is similar to injecting an extra process signal into the model. In both cases, the output error is made less sensitive to the disturbance. However, in injecting the model signal into the process, the model is left unaltered and so no concessions have to be made to the control goal of following the model. In this respect, this method is to be preferred over model adjustment. Note that in the example given in section 7.2, such a process correction was not possible because of the different points at which the disturbance acted on the process.

7.4 Summary

If the disturbance on a process directly depends on the reference signal, which is the case if the disturbance is generated by another process with the same reference, extra cross terms in the error equation arise. Because the disturbance has the largest effect at set-point changes, where the adaptation normally extracts most of its information, these terms inhibit a proper convergence of the controller parameters. If interaction between the processes is mutual, for each of the processes the interaction can be regarded as unmodelled dynamics. If both processes are controlled by an adaptive system the unmodelled dynamics are time varying. Hence, application of the decomposition method of chapter 5 is not straightforward.

Based on the same philosophy as was used in chapter 5, the reference model may be altered in such a way that the disturbance is also present in the reference model output, in a still desired fashion. The error equation is thereby made less sensitive to the disturbance, which may improve system behaviour and inhibit instability effects.

It should be stressed that the method presented in this chapter is only applicable if there is knowledge about the disturbance itself (either obtained directly, or derived from measurable signals), and about the effect of the disturbance on the process. This assumption is not required when applying a MIMO adaptive controller. However, at this time there is no formal solution for the MIMO adaptive control problem if only the outputs of the subsystems are known, and the existing results in this research area all need much *a priori* knowledge of the process. If all states are available the adaptive control scheme presented in section 2.2 is to be preferred, and it provides an exact solution. However, practical experiments have shown that model adjustment may provide an acceptable compromise in this area.

Problems

7.1 In chapter 5, model adjustment was presented as a method that yields a compromise between what is *desired* and what is *possible*. Do you recognize the same compromise in the use of model adjustment presented in this chapter?

7.2 Explain why the disturbance compensation injection into the *process*, as presented in the end of section 7.3, cannot be applied in the example given in section 7.2.

8

Epilogue

One of the aims of this book has been to give the reader a solid understanding of the mechanisms that play a role in adaptive control, thus giving him or her a sound basis for applying adaptive control in practice. For this reason, you have not found a great deal of mathematics in this book, but, more than that, you have found explanations why specific theoretical aspects are important and in which situations some mathematical requirements may *not* be needed in practice. For example, the mathematical description of the positive real property of linear systems is undoubtedly important. However, you need to understand the *reason* for the SPR requirement in order to understand, for example, why things go wrong if they do, and in what circumstances violation of the SPR requirement may be allowed. A similar reasoning is valid for other mathematical aspects.

What does the literature say?

Scanning the literature on adaptive control, the relatively large gap between theoretical and practical results is striking. The theory is normally based on strict assumptions regarding the process to be controlled. For example, knowledge of the relative degree of the process is needed and the process is required to be linear. In practice, these requirements are rarely met and therefore practical applications are usually implemented using a simple form of adaptive control. To ensure proper behaviour under all circumstances, heuristic solutions are often used to cope with practical problems, such as, for example, process input saturation. Although the theory is developing such that the strict assumptions lying at the basis of MRAC design are being increasingly relieved, the theoretical results are usually not in a form that can be made directly applicable. A better understanding can cure this problem.

Application area of MRAC

The main application area of MRAC is that of tracking problems, although some steps have been taken to take regulation behaviour into account (especially in discrete-time schemes). In practice, the design of a primary controller for a process of an order larger than, say, three, causes problems, especially if only the plant's input and output are available. Meeting the perfect model-matching demand requires a primary controller of high order, which usually generates a signal vector on the basis of derivatives of the process output. A high-order controller is generally more sensitive to disturbances, such as measurement noise or process nonlinearities, puts higher requirements on the reference signal r, and usually suffers from a slow convergence. In this book, the philosophy of adjusting the reference model output to the actual process and controller capabilities, taking into account the absence of the perfect model-matching property, has been presented. If there is some knowledge about the process dynamics, this approach increases the process degree allowed from about three to five or six.

The reference model

The importance of the reference model is usually underestimated (take note of the reference models used in the practical applications in this book). A general observation in this book is that the choice of a fixed reference model is not the best choice in all situations. On the one hand, it may not be possible to specify the desired behaviour by means of a fixed transfer function. For example, if the process suffers from input saturation, the process potential is not fully used if the reference model is chosen such that it can always be followed by the process. On the other hand, the capability of the controller to achieve perfect model matching may be limited, for example if the adaptive controller is of too low an order. In such a case, the actually achievable process response cannot be that of a fixed reference model. In both cases, model adjustment is a way of improving system behaviour. This model-adjusting strategy is shown in figure 8.1, which shows that

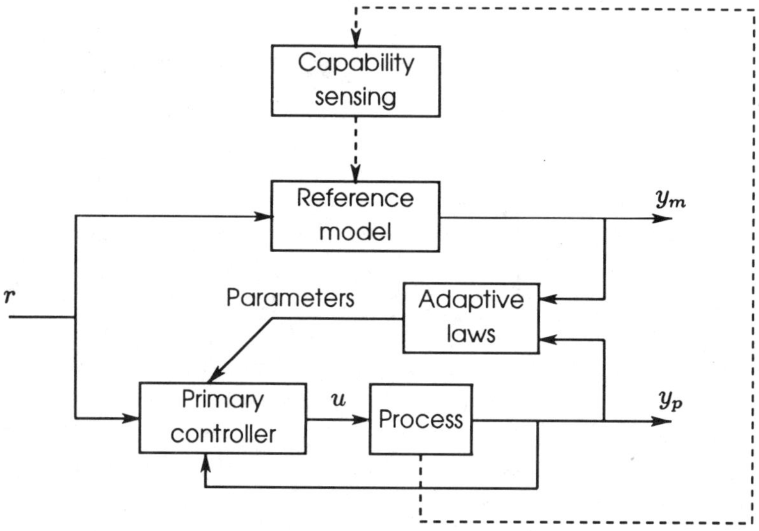

Figure 8.1 Reference model adjustment, based on 'capability sensing' of the process

the actual status is in some way extracted from the process, and the result of this 'sensing' is translated into an adjustment rule of the reference model. The sensing and translation can be done directly, as was the case in the chapters 5 and 7, but can also take place adaptively, as in chapter 6. In both cases, the reference model is altered such that its output yields a *desired* and an *achievable* response at the same time.

Error adjustment

Because the primary use of the reference model is to be able to calculate the error signal used in the adaptation, model adjustment can alternatively be regarded as *error adjustment*. Although these two are equivalent, the choice as to which description to use depends on which is easiest in a particular problem setting. The use of an orthogonal error signal, described in chapter 3 as a way of dealing with dead time in the process transfer function, is the easiest derived by inspecting the effect of the disturbance (the dead time) on the error signal. The orthogonal error is made insensitive to the disturbance in a heuristic manner.

Direct model adjustment

Unmodelled dynamics are process dynamics not accounted for in the controller. Although in the literature the term is usually used for *unknown* dynamics, the designer generally has a good idea of the process dynamics in the frequency range of interest. In the direct model adjustment presented in chapter 5, the aim was to include similar unmodelled dynamics in the reference model as present in the process. In this way, the error is made less sensitive to these dynamics, so that the adaptation does not react to disturbances that the primary controller cannot cancel. This direct model adjustment can only be performed with structural knowledge of the unmodelled process part, and some knowledge of the relevant parameter values. Hence, the direct model adjustment (reference model decomposition) is primarily a way of using a lower-order controller than necessary for perfect model matching. This is useful in a practical application, where a controller of low order is generally favoured. The division of the process dynamics into a nominal part, a structured unmodelled part, and a really unmodelled part indicates the difference from the usual robustness analysis regarding unmodelled dynamics.

A similar inclusion of a structural disturbance in the reference model was presented in chapter 7, where known external disturbances were considered. As before, the inclusion decreases the sensitivity of the output error with respect to the disturbance. The main application area of the method of chapter 7 is that of several processes influencing each other, and hence one is dealing with a multi-input, multi-output process. The interaction is considered to be a structural external disturbance for each of the processes. In the case of mutual interaction, the interaction takes the form of time-varying unmodelled dynamics for each process separately. It is therefore not surprising that a similar strategy as that discussed in chapter 5 can be applied.

Adaptive model adjustment

If the control specifications cannot be met by using a fixed reference model because the process capabilities are varying, adaptive adjustment of the model comes into play. This situation occurs mainly if the process puts severe limits on its input, which was also the case in the practical example of adaptive model adjustment given in chapter 6. Here, an approximately time-optimal motor behaviour was achieved by updating the reference model with the aid of a least-squares estimation of the load inertia. The desired time-optimal response depended on the unknown value of this inertia, making the use of a fixed reference model impossible.

Why use model adjustment?

The general way of making the error signal insensitive to some disturbance, either by means of model adjustment or by means of (equivalent) error modification, has proved successful. Note that the reason for requiring such a strategy was, in all cases, violation of the theoretical assumptions. The process was either not linear, was of an order higher than was accounted for in the controller, was disturbed by a parallel process, or had dead time in its transfer function. Also, in the case of the direct-drive motor, the desired closed-loop behaviour was not constant but depended on some of the unknown process parameters.

Is model adjustment sensible?

As stated, the main observation in this book is that the reference model, specifying the control goal, should in some cases be made dependent on the process capabilities. One may ask whether automatic adjustment of the *control goal* is sensible. The answer is that in the case of MRAC such an adjustment may be *essential*. The model adjustment is a compromise between the user-specified reference model on the one hand and the limited capabilities of the primary controller to obtain a matching closed-loop response on the other. It is, however, clear that the designer should use this tool with care because the behaviour that actually will occur cannot be completely anticipated beforehand.

What should you have learned from this book?

Hopefully, this book will fill a small part in the building of an important structure: the bridge between theory and practice in adaptive control. In the author's opinion, the capability of rating the true value of specific theoretical results bridges half the gap. For this, *understanding* adaptive control instead of only being able to write down relevant formulas is a necessary prerequisite. This book has given you enough background for such an understanding. So who will bridge the other half of the gap between theory and practice? The answer is simple: *you* will. In applying adaptive control yourself, you will learn many things that no book can teach you...

A

Stability theory

This appendix is a compact introduction to Lyapunov stability and passivity and positivity concepts. Both play an important role in adaptive systems. The interested reader is refered to (Slotine and Li, 1991) for a more elaborate description.

A.1 Lyapunov stability

Consider the nonlinear, time-varying system:

$$\dot{\boldsymbol{x}} = \boldsymbol{f}(\boldsymbol{x}, t) \tag{A.1}$$

Here, $\boldsymbol{f}$ is an $n \times 1$ vector function and $\boldsymbol{x}$ is the $n \times 1$ state vector. While the above system is time dependent, it is termed *nonautonomous*. A time-invariant system is termed *autonomous*. In adaptive control, the closed-loop process on its own can be considered a nonautonomous linear system (because the parameter vector $\boldsymbol{\theta}$ varies with time), while the complete system including the adaptive laws can be regarded as an autonomous nonlinear system. Hence, in the following only autonomous nonlinear systems will be considered. An *equilibrium point* $\boldsymbol{x}^*$ of system (A.1) is characterized by $\boldsymbol{f}(\boldsymbol{x}^*) = 0$. If the state $\boldsymbol{x}$ of system (A.1) is situated at an equilibrium point at $t = 0$, it will stay on the equilibrium point for all $t > 0$.

Stability of an equilibrium point is mainly concerned with the question of what happens if the state vector at time $t = 0$ deviates from the equilibrium state: $||\boldsymbol{x}(0)|| = ||\boldsymbol{x}^*|| + \delta$. In other words, if the initial state is not exactly equal to the equilibrium state, what will happen? Basically, there are four possibilities.

1. The state $\boldsymbol{x}$ can be made to remain in any specified vicinity ϵ of $\boldsymbol{x}^*$ by choosing a sufficiently small deviation δ of $\boldsymbol{x}^*$ at $t = 0$. However, $\boldsymbol{x}$ is not guaranteed to *converge* to $\boldsymbol{x}^*$. This behaviour is called *stable (S)*.

2. The system is not stable. It is then said to be *unstable*. Note that if a system is unstable, this does not imply divergence from $\boldsymbol{x}^*$. It merely states that it is possible to specify an ϵ which does not allow any δ. A well-known example of this is a harmonic oscillator which is oscillating with a specific frequency and amplitude: $x(t) = A \cdot \sin(\omega t)$. While x always lies between $-A$ and $+A$, specifying $\epsilon > A$ allows any initial deviation $\delta < \epsilon$. However, no value of δ can be found that allows an $\epsilon < A$ because x will always exceed ϵ for some t.

3. The state $\boldsymbol{x}$ will converge to $\boldsymbol{x}^*$ as long as the initial deviation δ of $\boldsymbol{x}^*$ is smaller than some boundary R. More formally, if $\delta < R$, for an arbitrarily close vicinity μ of $\boldsymbol{x}^*$ a time t^* exists such that for all $t > t^*$, $||\boldsymbol{x} - \boldsymbol{x}^*|| < \mu$. This notion is called *asymptotic stability (AS)*. Note that in this definition, starting within a boundary R from $\boldsymbol{x}^*$ does not imply that $\boldsymbol{x}$ will remain within R for all $t > 0$, but only that $||\boldsymbol{x} - \boldsymbol{x}^*|| \to 0$ as $t \to \infty$.

4. If the asymptotic stability is guaranteed for any inital deviation δ, the asymptotic property is said to be *global (GAS)*.

To check to which stability class a given system of the form (A.1) belongs, Lyapunov's direct method can be applied. Lyapunov's method can only investigate stability properties of an equilibrium point $\boldsymbol{x}^* = 0$. Hence, other equilibrium points have to be transformed to $\boldsymbol{x}^* = 0$ by a transformation $\boldsymbol{x}' = \boldsymbol{x} - \boldsymbol{x}^*$. In applying Lyapunov's method, first a Lyaponov function $V(\boldsymbol{x})$ is defined. This function can be considered as a sort of *energy function*, while it has similar properties to the energy stored in the system. The Lyapunov function itself must satisfy:

$$V(\boldsymbol{x}) > 0, \quad \forall\, \boldsymbol{x} \neq 0$$

In addition, to be comparable to an energy function, $V(\boldsymbol{x})$ should be monotonically increasing, and go to infinity as $||\boldsymbol{x}|| \to \infty$:

$$0 < \alpha||\boldsymbol{x}|| < V(\boldsymbol{x}); \quad \alpha > 0$$

$$V(\boldsymbol{x}) \to \infty \ \text{ as } \ ||\boldsymbol{x}|| \to \infty$$

Now, it can be felt that if the stored energy in a system decreases as time passes, all energy will eventually leave the system and the equilibrium $\boldsymbol{x} = 0$ will be reached. Similarly, if the time derivative of the Lyaponov function is always negative and hence V is decreasing with time, V will eventually become zero because V is monotonous. As $V = 0$ implies $\boldsymbol{x} = 0$, also due to the monotonous character of V, a negative definite $\dot{V}$ guarantees asymptotic stability:

$$\dot{V} < 0, \quad \forall\, \boldsymbol{x} \neq 0$$

To calculate $\dot{V}$, partial derivatives of V with respect to the elements of $\boldsymbol{x}$ are needed:

$$\dot{V}(\boldsymbol{x}, t) = \frac{\mathrm{d}\boldsymbol{x}}{\mathrm{d}t} \cdot \frac{\partial V}{\partial \boldsymbol{x}} = \boldsymbol{f}(\boldsymbol{x}) \cdot \frac{\partial V}{\partial \boldsymbol{x}}$$

Hence, the partial derivatives $\partial V / \partial \boldsymbol{x}$ must be continuous. The above requirements on V guarantee global asymptotic stability (GAS). By using less strict requirements, other forms of stability are obtained. For example, if $\dot{V}$ is negative *semi*-definite, which implies that $\dot{V} = 0$ for some $\boldsymbol{x} \neq 0$, the stability is no longer asymptotic. If V is not monotonuously increasing with $||\boldsymbol{x}||$ but just increasing, and V does not go to infinity for $||\boldsymbol{x}|| \to \infty$, the stability is not global. Note that Lyapunov's method provides a stability *guarantee*: if the requirements mentioned are met, the system is guaranteed to be (globally asymptotically) stable. However, if the requirements are not satisfied, the system may still be stable. In addition, the choice of the Lyapunov function V is crucial in the stability check. Different Lyapunov functions may give different stability results.

For a linear system $\dot{\boldsymbol{x}} = \boldsymbol{A}\boldsymbol{x}$, let us consider a quadratic Lyapunov function: $V = \boldsymbol{x}^T \boldsymbol{P} \boldsymbol{x}$, in which $\boldsymbol{P}$ is a symmetric positive definite matrix. Then:

$$\dot{V} = \dot{\boldsymbol{x}}^T \boldsymbol{P} \boldsymbol{x} + \boldsymbol{x}^T \boldsymbol{P} \dot{\boldsymbol{x}} = -\boldsymbol{x}^T \boldsymbol{Q} \boldsymbol{x}$$

in which:

$$-Q = A^T P + PA$$

According to Lyapunov's theorem, a positive definite symmetric Q always yields a positive definite symmetric matrix P if the system $\dot{x} = Ax$ is asymptotically stable.

A.2 Positivity and passivity

Positivity and passivity are two equivalent notions. Positivity is a property of linear systems, while passivity is a more general concept that is applicable to both linear and nonlinear systems. A scalar, controllable, linear system with input u and output y:

$$\begin{aligned} \dot{x} &= Ax + bu \\ y &= c^T x \end{aligned} \qquad \text{(A.2)}$$

with a transfer function:

$$H(s) = \frac{Y(s)}{U(s)} = c^T (sI - A)^{-1} b$$

is said to be *positive real* (PR) if $\text{Re}[H(s)] \geq 0$ for all $\text{Re}[s] \geq 0$. Hence, the real part of the transfer function can never become negative as long as the real part of s is larger than or equal to zero. For example, if $s = j\omega$, the Nyquist diagram of $H(j\omega)$ must lie in the right half of the complex plane, including the imaginary axis. This implies that the phase shift of $H(s)$ must be between $-90°$ and $+90°$ (both inclusive). $H(s)$ is *strictly positive real* (SPR) if $\text{Re}[H(s)] > 0$ for all $\text{Re}[s] \geq 0$, excluding the possibility that $\text{Re}[H(s)] = 0$. The phase shift allowed now excludes $-90°$ and $+90°$.

To check if a given transfer function is PR or SPR, the Kalman–Yakubovich lemma is a useful tool. The system (A.2) is strictly positive real if there exist positive definite matrices P and Q such that:

$$\begin{aligned} A^T P + PA &= -Q \\ Pb &= c \end{aligned} \qquad \text{(A.3)}$$

Hence, if system (A.2) is asymptotically stable, and in addition $\boldsymbol{b}$ and $\boldsymbol{c}$ are related according to (A.3), the system is SPR. For positive real systems, the existence of a semi-negative matrix $\boldsymbol{Q}$ in (A.3) is sufficient.

A more complex form of equation (A.3) exists called the *Meyer–Kalman–Yakubovich lemma* (Slotine and Li, 1991). This lemma differs from the Kalman–Yakubovich lemma in that only stabilizability of system (A.2) is required instead of controllability. In adaptive control, $\boldsymbol{A}$ usually belongs to the designer-chosen reference model and hence controllability can easily be guaranteed. The use of the much-simpler Kalman–Yakubovich lemma is therefore propagated.

Passivity is a generalized form of positivity. A linear system that is passive must be positive real, and the reverse is also true. Consider a system with input $\boldsymbol{u}$ and output $\boldsymbol{y}$. The energy in the system depends on the external input of power and on the power generation in the system:

$$\frac{\mathrm{d}}{\mathrm{d}t}\,[\text{Stored energy}] = [\text{ext. power input}] + [\text{int. power generation}]$$

Considering $\boldsymbol{u}$, the input, as 'voltage' (or, generally, as effort variable), and $\boldsymbol{y}$, the output, as 'current' (or, generally, as flow variable), the external power input equals $\boldsymbol{u}^T\boldsymbol{y}$. If the internal power generation is negative, the system is said to be *dissipative* or *strictly passive*. If the internal power generation is less than or equal to zero, the system is *passive*. Roughly speaking, strict passivity is equivalent to SPR and asymptotic stability; passivity is equivalent to PR and stability. The definition of passivity is generally given as:

$$\exists\, \chi < \infty : \quad \int_{\tau=0}^{t} \boldsymbol{u}^T\boldsymbol{y}\,\mathrm{d}\tau \geq -\chi \quad \forall\, t > 0$$

The main result using positivity and passivity concepts is that any parallel combination of passive blocks is also passive. A feedback combination of two passive blocks in which at least one is strictly passive is strictly passive. This is of great interest in hyperstability theory, in which an SPR (and hence, strictly passive) linear block is connected to a passive nonlinear block in a feedback configuration. This combination is strictly passive (and, hence, asymptotically stable).

B

Answers to Problems

Chapter 1

1. Adaptive schemes that are guaranteed to be stable usually are so for any positive adaptation gain matrix $\boldsymbol{\Gamma}$. The adaptation rate is then of little importance, as long as its sign is correct. In schemes that do not allow for a stability proof, as for example the sensitivity method, too large an adaptation gain usually induces stability problems. A large error signal and signal vector have a larger 'effective adaptation gain', and hence are more prone to yield instability than are small signals. To prevent such an effect, especially in those schemes for which a stability proof cannot be given, normalization is useful.

2. If the plant is of first order, the adaptive controller design is much simpler than that for higher-order plants, because of:

   (a) the simplicity of the primary controller. A representation of the complete state vector (consisting of only one state, the output) is available and hence no state-reconstruction or other complex primary control method need to be used. The reference model is of first order and so the perfect model-matching condition is easily satisfied.

(b) the reference model being of order one, the linear part of the error equation for the output error $e = y_p - y_m$ is always SPR. This implies that e can be used directly in the adaptation without modification.

3. In the derivations, it is often assumed that $\dot{\boldsymbol{\phi}} = \dot{\boldsymbol{\theta}}$, which is usually true while $\boldsymbol{\phi} = \boldsymbol{\theta} - \boldsymbol{\theta}^*$, with $\boldsymbol{\theta}^*$ the parameter vector for which $y_p = y_m$. Now, if the process parameters are not constant, $\boldsymbol{\theta}^*$ is no longer constant, and hence $\dot{\boldsymbol{\phi}} \neq \dot{\boldsymbol{\theta}}$.

4. There are two reasons why both $\boldsymbol{e}$ and $\boldsymbol{\phi}$ are present in the Lyapunov function. First, Lyapunov's method is used to consider stability of the complete adaptive system, and hence all states must be present in the Lyapunov function. Any state not included may be unstable without Lyapunov's method detecting such a situation. Due to the adaptation, $\boldsymbol{\phi}$ should also be considered as a set of states. Second, the derivation of the adaptive laws is directly based on putting to zero some of the terms in $\dot{V}$. If $\boldsymbol{\phi}$ were not included in V, such a procedure would be impossible.

Note that Lyapunov's method can only provide a stability result for an equilibrium of 0 of all states included. Therefore, for example $\boldsymbol{\theta}$ cannot be included in V, for all states included must be desired to go to zero. This is true for both $\boldsymbol{e}$ and $\boldsymbol{\phi}$.

5. In Lyapunov's method, an error signal $\epsilon = \boldsymbol{p}^T\boldsymbol{e}$ is often used. To see how this is connected to hyperstability properties, consider a state-space description of the error equation:

$$\begin{aligned} \dot{\boldsymbol{e}} &= \boldsymbol{A}_m\boldsymbol{e} + \boldsymbol{b}(-w) \\ \epsilon &= \boldsymbol{c}^T\boldsymbol{e} \end{aligned}$$

Here, w is the output of the nonlinear part of the error equation, which is usually equal to $-\boldsymbol{\phi}^T\boldsymbol{\omega}$. The transfer function $\epsilon/(-w)$ is SPR if, according to the Kalman–Yakubovich lemma:

$$\begin{aligned} \boldsymbol{A}_m^T\boldsymbol{P} + \boldsymbol{P}\boldsymbol{A}_m &= -\boldsymbol{Q} \\ \boldsymbol{P}\boldsymbol{b} &= \boldsymbol{c} \end{aligned}$$

The first equation is always satisfied because $\boldsymbol{A}_m$ belongs to an asymptotically stable system (namely, the reference model). The second equation relates the observation vector $\boldsymbol{c}$, to be chosen such that $\epsilon/(-w)$ is SPR, to the matrix $\boldsymbol{P}$. If the error equation is in phase-variable form, and hence $\boldsymbol{b}^T = (0, 0, \ldots, 0, 1)$, $\boldsymbol{c}$ is chosen equal to the last column of $\boldsymbol{P}$.

In the given adaptation error, $\epsilon = \boldsymbol{p}^T\boldsymbol{e}$, $\boldsymbol{p}$ is the observation vector of the system producing $\boldsymbol{e}$. Hence, if $\boldsymbol{p}$ is the last column of $\boldsymbol{P}$, $\boldsymbol{p}$ automatically makes the error equation SPR if the error system is in phase-variable form. This appears to be the case in general, and hence the link between Lyapunov's method and the hyperstability approach has been laid.

6. The complete adaptive law becomes:

$$K(t) = -\gamma' \int_{\tau=0}^{t} er \, d\tau - \beta' er$$

Hence, Popov's inequality extends to:

$$\begin{aligned}
\int_{t=0}^{T} \epsilon w \, dt &= -\int_{t=0}^{T} e\,(bK - 1)\, r \, dt \\
&= -\int_{t=0}^{T} er \left\{ b \left[\int_{\tau=0}^{t} -\frac{\gamma}{b} er \, d\tau + K(0) - \frac{\beta}{b} er \right] - 1 \right\} dt \\
&= \int_{t=0}^{T} er \, \gamma \left[\int_{\tau=0}^{t} er \, d\tau - \frac{b}{\gamma} K(0) + \frac{1}{\gamma} \right] dt \\
&\quad + \int_{t=0}^{T} er \, \beta \, er \, dt
\end{aligned} \tag{B.1}$$

The first integral in (B.1) is equal to that given in example 1.3, and passes Popov's test. Note that while the proportional adaptation is presented as an *addition* to the integral adaptation, γ is nonzero in example 1.3. Hence, the term containing the initial condition $-\frac{b}{\gamma}K(0) + \frac{1}{\gamma}$ is dealt with by the integral adaptation. The second integral in (B.1) is easy to check:

$$\int_{t=0}^{T} er \, \beta \, er \, dt = \beta \int_{t=0}^{T} (er)^2 \, dt \geq 0$$

This shows that proportional adaptation, as an addition to integral adaptation, does not violate the stability properties. In fact, proportional adaptation by itself does not do so either, but violates the *asymptotic* properties of the stability. While hyperstability only gives conditions for asymptotic stability, it cannot be used in this case. Try for yourself to use hyperstability for proportional adaptation without an integrating term, and you will see that the initial value of ϕ will give you some trouble.

The 'switch-like' addition to the adaptive law:

$$K(t) = -\gamma' \int_{\tau=0}^{T} er \, d\tau - \rho' \text{sign}(er)$$

yields a similar result in that the first integral is equal to that given in (B.1), and the second one changes to:

$$\rho \int_{t=0}^{T} er \, \text{sign}(er) \, dt$$

which is always larger than 0.

Chapter 2

1. As already mentioned in chapter 1, the effects that may be encountered by using integral adaptation are similar to those that occur in nonadaptive control, where it can be seen that a pure integrating term in the primary controller may lead to oscillations or unstable behaviour. In MRAC, a stability proof usually guarantees the stability, but large adaptation gains may still give rise to oscillations in the controller parameters. A large adaptation gain may prohibit a fast convergence, as in nonadaptive control an integrating part in the controller may result in overshoot or badly damped behaviour.

2. In the augmented error method L occurs at two points: as filters L^{-1} in the error-augmenting network which also produce $\boldsymbol{\xi}$ from $\boldsymbol{\omega}$, and integrated in the reference model transfer in $W_m L$. The filters L^{-1} incur no problems because their relative degree is larger than zero. However, an order of $n-m+1$ of L would give $W_m L$ a relative degree of -1, which implies that $W_m L$ has one more zero than poles. Hence, $W_m L$ is not implementable.

3. In this specific simulation, the process and model transfers were:

$$W_p = \frac{12}{s^2 + 2s + 8}; \quad W_m = \frac{16}{s^2 + 8s + 16}$$

Hence,

$$Z_p = 1, \quad R_p = s^2 + 2s + 8, \quad k_p = 12$$
$$Z_m = 1, \quad R_m = s^2 + 8s + 16, \quad k_m = 16$$

In addition, the following holds for the ASG denominator polynomial N and the controller polynomials C and D:

$$N = s + 10, \quad C = c_1, \quad D = d_1$$

Substituting all these elements in equation (2.27), the closed-loop transfer function becomes:

$$\begin{aligned}\frac{y_p}{r} &= \frac{k_0 \times 12 \times 1 \times (s+10)}{(s+10-c_1)(s^2+2s+8)-(d_0 s+10d_0+d_1)\times 12 \times 1} \\ &= \frac{12k_0(s+10)}{s^3+(12-c_1)s^2+(28-2c_1-12d_0)s+(80-8c_1-120d_0-12d_1)}\end{aligned}$$

To make this closed loop equal to the reference model, the factor $(s+10)$ in the numerator must be cancelled by the denominator. By applying long division of the denominator by $(s+10)$, the dividend becomes: $s^2+(2-c_1)s+(8+8c_1-12d_0)$, and the remainder is: $-88c_1-12d_1$. By putting the remainder equal to zero, and by stating the dividend equal to the reference model transfer function denominator $s^2+8s+16$, the following controller parameters emerge:

$$\begin{aligned} c_1^* &= -6 \\ d_0^* &= -4\frac{2}{3} \\ d_1^* &= 44 \\ k_0^* &= \frac{4}{3} \end{aligned}$$

The remaining term in the numerator, $12k_0$, should be equal to the corresponding factor 16 in the reference model transfer function, while the denominator of y_p/r is monic. This leads to the value of $k_0^* = 4/3$.

4. The transfer functions of the closed-loop process and the reference model are, respectively:

$$\frac{y_p}{r} = \frac{2k_p}{s^2+8s+2k_p}; \quad \frac{y_m}{r} = \frac{16}{s^2+8s+16}$$

Hence, the differential equations of y_p and y_m are:

$$\begin{aligned} \ddot{y}_p + 8\dot{y}_p + 2k_p y_p &= 2k_p r \\ \ddot{y}_m + 8\dot{y}_m + 16y_m &= 16r \end{aligned}$$

and so the error equation becomes:

$$\ddot{e} + 8\dot{e} + 16e = (2k_p - 16)(r - y_p)$$

Alternatively written, e is equal to:

$$e = \frac{2}{s^2 + 8s + 16}(k_p - 8)(r - y_p)$$

Now, the linear part equals $H = 2/(s^2 + 8s + 16)$, which is not SPR. An error-augmenting network is therefore used, which is designed along the same lines as given in example 2.5. Here, a polynomial $L = s + 4$ is used which makes HL SPR:

$$\begin{aligned} \epsilon &= e + HL\left(\boldsymbol{\theta} L^{-1} - L^{-1}\boldsymbol{\theta}\right)^T \boldsymbol{\omega} \\ &= e + \frac{2}{s+4}\left[k_p \frac{1}{s+4}(r - y_p) - \frac{1}{s+4} k_p (r - y_p)\right] \end{aligned}$$

Chapter 3

1. The step signal has a degree of PE of one, as noted in example 3.1. While there are three parameters, this signal is obviously insufficient to yield parameter convergence. The sine wave has two spectral lines (one for $-\omega$ and one for $+\omega$), and hence has a degree of PE of 2. Intuitively, this can be explained by considering that a sine function is determined by two aspects, namely its amplitude and its phase. The sine wave by itself is not sufficient to assure convergence. However, addition of the two signals yields a degree of PE of 3 which guarantees convergence of $\boldsymbol{\theta}$ to $\boldsymbol{\theta}^*$.

2. No. In γ-modification, the idea is that as the adaptation error ϵ becomes smaller, the modification plays a lesser and lesser role. A sufficiently large degree of PE of the reference signal makes sure that ϵ actually decreases and is not increased by a possible leaking effect. As ϵ tends to zero, the modification is switched off and the adaptation approaches the normal integral adaptation. In σ-modification, however, leakage always occurs, and the parameters will not converge to their true values. This effect is solely dependent on the value of σ and cannot be compensated by a high degree of PE of the reference signal.

3. Dynamically altering the dead zone magnitude as a function of signals in the adaptive system is not straightforward. It would not be useful to base such a strategy solely on $|\epsilon|$, while the disturbance in ϵ cannot be distinguished from the actual process-model error. In the literature (Kreisselmeier and Anderson, 1986), a method is presented in which the dead zone magnitude is made dependent on the process input and output. A large degree of PE of these signals increases the dead zone magnitude. However, this method is specifically meant for disturbances where the magnitude depends on the signal vector, which is true for unmodelled dynamics but not for external disturbances. Hence, on-line changing of the dead zone magnitude is not regarded as the most usual thing to do.

4. To cancel the effect of δ on the output y, x_2 must have a ramp-like signal that eventually, after adaptation, becomes equal to $-\delta$. However, the reference model state that corresponds to x_2 does not have such a compensation term, and hence the difference between x_2 and the corresponding model state must be nonzero to obtain $y = y_m$. Because in the method using state feedback a 'compensated error' $\boldsymbol{p}^T\boldsymbol{e}$ is used that includes all state errors, to achieve perfect model matching $\boldsymbol{p}^T\boldsymbol{e}$ must be nonzero. Because a nonzero $\boldsymbol{p}^T\boldsymbol{e}$ invokes the adaptation mechanism, the parameters will be updated even if $y = y_m$.

   The augmented error method only makes use of input and output measurements, and hence the error used in the adaptation will actually become zero if $y = y_m$. The adaptive compensation is then readily implemented.

Chapter 4

1. The one-sample delay actually makes the linear part of the error equation non-SPR, as does a time delay in continuous time. As mentioned in chapter 1, particularly for MRAC schemes for which a stability proof cannot be given, normalization is useful. Because a non-SPR error equation violates the stability requirements, normalization would be useful in discrete MRAC for the same reasons. Fortunately, the *a priori / a posteriori* correction actually restores the stability properties which where violated by the extra time delay, which is not generally true for the normalization in continuous case.

2. In terms of the modified hyperstability theory, the linear part of the error equation is of class $L(0.5)$, while $W_m L$ may be shifted over a length of 0.25 before the linear part becomes non-SPR. Hence, in the least-squares method (equation (4.10)), λ_2 may have a maximum value of 0.5, while this

method is of class $N(\lambda_2)$. The 'standard' least-squares method has $\lambda_2 = 1$, and hence does not meet the requirements put by the modified hyperstability theory. Note that the maximum allowed value for λ_2, i.e. 0.5, implies a parameter update mechanism somewhere between least-squares and gradient adaptation.

3. The deadbeat MRAC scheme is of an indirect type, while the process parameters are explicitely estimated, and this estimation is used in the controller calculation. However, it is still possible to analyze the complete system as a whole, including the adaptation. This is possible only for a very limited class of indirect adaptive schemes.

Chapter 5

1. Attempting model matching in a frequency range where the unmodelled dynamics play an important role is dangerous because to achieve such matching the plant will have to be excited in this high-frequency range. In other words, the control signal will contain the high frequencies needed for perfect model matching, which will, however, also excite the unmodelled dynamics. Hence, the plant's output will be considerably affected by the unmodelled dynamics, which can therefore easily penetrate into the adaptation mechanism.

   In this respect, a 'fast' reference model is obviously more dangerous than a 'slow' one. Another way of looking at it is to consider the parameter values needed for perfect model matching. A high closed-loop bandwidth requires a 'tight' feedback, which may easily induce problems if unmodelled dynamics are present.

2. With the given control, $k_0^* = 4$ and $d_0^* = -3$, as illustrated in example 5.2. The actual closed loop with the full-order process is:

$$\frac{y_p}{r} = \frac{10k_0^*}{s^2 + 11s + 10 - 10d_0^*} = \frac{40}{s^2 + 11s + 40}$$

which is equal to $\overline{W}_m$. $\widetilde{W}_m$ can be calculated by subtracting W_m:

$$\begin{aligned}\widetilde{W}_m &= \overline{W}_m - W_m \\ &= \frac{40}{s^2+11s+40} - \frac{4}{s+4} \\ &= -4\frac{s^2+s}{s^3+15s^2+84s+160}\end{aligned}$$

3. The full-order process is given by:

$$\overline{W}_m = \frac{1}{s+1}\cdot\frac{10}{s+10}$$

and hence the unmodelled part equals $10/(s+10)$. This immediately gives us the decomposition:

$$\frac{10}{s+10} = \frac{T_1}{N-T_2}$$

Now $T_1 = 10$ and $N - T_2 = s + 10$, which allows freedom in choosing N.

4. The original disturbance equals:

$$\widetilde{W}_m r = -4\frac{s^2+s}{s^3+15s^2+84s+160}\, r$$

The disturbance using decomposition becomes:

$$\overline{W}_m r - y_m^* = -\frac{T_{p1}}{N}\widetilde{W}_p\left(\theta^{*T}\omega\right) + \frac{T_1}{N}\widetilde{W}_m r$$

as described in example 5.5. Applying the given decomposition with $T_2 = 0$:

$$T_1 = T_{p1} = 10; \quad N = N_p = s + 10$$

which yields:

$$\overline{W}_m r - y_m^* = \left(-\frac{10}{s+10}\right)\widetilde{W}_p\left(\theta^{*T}\omega\right) + \frac{-40}{s+10}\cdot\frac{s^2+s}{s^3+15s^2+84s+160}$$

As in example 5.5:

$$\widetilde{W}_p\left(\boldsymbol{\theta}^{*T}\boldsymbol{\omega}\right) = \left(\frac{-N_p + T_{p1} + T_{p2}}{T_{p1}}\right)\overline{W}_m r$$

$$= \left(-\frac{s}{10}\right)\overline{W}_m r$$

And hence:

$$\overline{W}_m r - y_m^* = \frac{-10}{s+10}\cdot\frac{-s}{10}\cdot\frac{40}{s^2+11s+40}r$$

$$-\frac{40}{s+10}\cdot\frac{s^2+s}{s^3+15s^2+84s+160}r$$

$$= \frac{120s}{s^4+25s^3+234s^2+1000s+1600}r$$

which is the disturbance after decomposition. Note that if the decomposition had been chosen in a more complex fashion ($T_2 \neq 0$), which is usually the case, the expression for the disturbance would have been more complex. While the decomposition parameters have not been chosen sensibly it is not very useful to investigate properties of the new disturbance with respect to the original one.

Chapter 6

First, note that because $a_{p1} = a_{m1}$, the perfect model-matching condition is still met only if a_{p2} is adjusted. Using only $e_2 = x_{p2} - x_{m2}$ in the adaptive law:

$$\dot{a}_{p2} = +\gamma_2 e_2 x_{p2}$$

actually means that the compensator vector $\boldsymbol{p}$ equals $\begin{pmatrix} 0 \\ 1 \end{pmatrix}$. Hence:

$$\boldsymbol{P} = \begin{pmatrix} p_{11} & 0 \\ 0 & 1 \end{pmatrix}$$

Taking the derivative of the standard Lyapunov function:

$$V = \boldsymbol{e}^T \boldsymbol{P} \boldsymbol{e} + \boldsymbol{\phi}^T \boldsymbol{\Gamma}^{-1} \boldsymbol{\phi}$$

yields:

$$\dot{V} = \boldsymbol{e}^T \left(\boldsymbol{A}_m^T \boldsymbol{P} + \boldsymbol{P} \boldsymbol{A}_m \right) \boldsymbol{e} + \{ \text{ some terms including } \boldsymbol{\phi} \ \}$$

As before, the extra terms including $\boldsymbol{\phi}$ are put to zero to obtain the adaptive laws. Note that in this specific case $a_{p1} = a_{m1}$ and hence the first element of $\boldsymbol{\phi}$ is already zero. No adjustment of a_{p1} is therefore necessary. In the first term:

$$\begin{aligned} \boldsymbol{e}^T \left(\boldsymbol{A}_m^T \boldsymbol{P} + \boldsymbol{P} \boldsymbol{A}_m \right) \boldsymbol{e} &= \boldsymbol{e}^T \begin{pmatrix} 0 & p_{11} - a_{m1} \\ p_{11} - a_{m1} & -2a_{m2} \end{pmatrix} \boldsymbol{e} \\ &= -\boldsymbol{e}^T \boldsymbol{Q} \boldsymbol{e} \end{aligned}$$

Hence, $\boldsymbol{Q}$ is not positive definite, but only positive. More specifically, the derivative of V is negative definite in e_2, but only negative semi-definite in e_1. Hence, e_2 is guaranteed to approach zero, but e_1 is not. This contradicts the statement that using only e_2 in the adaptive law is sufficient. Fortunately, stability theory proves conservative in this case, as it can easily be seen that for $t \to \infty$, $e_1 \to 0$. Because for $t \to \infty$, $x_{p2} \to x_{m2}$, a possible error between x_{p1} and x_{m1} must approach a constant value. While the DC gain of the model equals that of the process (both gains are 1) the error between x_{p1} and x_{m1} must also approach zero.

Chapter 7

1. Yes. Matching the reference model response is, of course, desired. However, within the limitations of the too-simple primary controller, which cannot properly cope with the interaction, this desired response can never be obtained. By letting the reference model have a similar 'disturbance' at its output as is present in the actual process, the inevitable error that occurs at the process output due to the interaction does not result in a large error signal. Hence, the adaptation will not react to disturbances that cannot be cancelled by the primary controller. This is exactly equivalent to the use of model adjustment in the case of unmodelled dynamics.

2. Injection into the process of a signal that will cancel the expected disturbance is not possible in the example given in section 7.2. The disturbance at the

output of process 2 is generated by a first-order plant (process 1). Hence, a compensating signal for process 2 (which is of second order) would require at least one differentiation of the reference r, which is not possible even without adaptation. In the practical example of the propeller setup, the disturbance due to the interaction acts on the same position in the system as does the control u (at least, there are no dynamics between these two). In such a case, the proposed injection *is* possible.

Bibliography

Ackermann, J. (1986): 'Robuste Regelung: Beispiele – Parameterraum – Verfahren', *Proc. Aussprachetag Robuste Regelung, Langen, BRD*, pp. 1–18.

Alonso, M. and E.J. Finn (1980): *Fundamental University Physics, part I*, Addison-Wesley, Reading, Massachusetts.

Amerongen, J. van (1982): Adaptive Steering of Ships: A model-reference approach to improved manoeuvring and economical course keeping, Ph.D. thesis, Control Laboratory, Delft University of Technology.

Amerongen, J. van and G. Honderd (1983): 'Design of model-reference adaptive systems – a comparison of the stability and the sensitivity approach', *Proc. IFAC Workshop on Adaptive Control and Signal Processing*, San Fransisco, USA.

Åström, K.J. (1983): 'Analysis of Rohrs counterexamples to adaptive control', *Proc. 22nd IEEE Conf. on Decision and Control*, San Antonio, pp. 982–987.

Åström, K.J and B. Wittenmark (1984): *Computer Controlled Systems: Theory and design*, Prentice-Hall, Englewood Cliffs.

Åström, K.J and B. Wittenmark (1989): *Adaptive Control*, Addison-Wesley, Reading, Massachusetts.

Bitmead, R. (1984): 'Persistence of excitation conditions and the convergence of adaptive schemes', *IEEE Trans. Inf. Theory*, **IT-30**

Bosch, P.P.J. van den (1987): 'Model reference adaptive control: Model updating'. In: *Systems & Control Encyclopedia*, pp. 3070–3071, Pergamon Press, Oxford.

Butler, H., G. Honderd and J. van Amerongen (1989): 'Model reference adaptive control of a direct-drive DC motor', *IEEE Control Systems Magazine*, **9**, pp. 80–84.

Butler, H. (1990): 'Model reference adaptive control', *Journal A*, **31** no.2, pp. 25–30.

Butler, H., G. Honderd and J. van Amerongen (1990a): 'Structured unmodelled dynamics in model reference adaptive control', *Proc. Sixth Yale Workshop on Adaptive and Learning Systems*, New Haven, USA, 15–17 August.

Butler, H., G. Honderd and A.R.M. Soeterboek (1990b): 'Ripple-free model reference adaptive control', *Proc. 29th IEEE Conf. on decision and control*, Honolulu, Hawaii, 5–7 December.

Butler, H., G. Honderd and J. van Amerongen (1991a): 'Model reference adaptive control of a gantry crane scale model', *IEEE Control Systems Magazine*, **11**, pp. 57–62.

Butler, H., G. Honderd and J. van Amerongen (1991b): 'Reference model decomposition in direct adaptive control', *International Journal of Adaptive Control and Signal Processing*, **5**, pp. 199–217.

Butler, H. and A.R.M. Soeterboek (1991): 'Model reference adaptive control of coupled systems', *Proc. IMACS–IFAC Symposium Modelling and Control of Technological Systems*, Lille, France, pp. 3–8.

Chiang, C.C. and B.S. Chen (1987): 'Imperfect model reference adaptive control: Deadbeat output error approach', *Int. J. of Control*, **46**, pp. 2011–2025.

Draijer, W. (1988): Adaptive Control: Theory and application, Master's thesis Lab. of measurement and control, Delft University of Technology.

Egardt, B. (1980): 'Unification of some discrete-time adaptive control schemes', *IEEE Trans. Aut. Control*, **AC–25**, pp. 693–697.

Ten Hengel, K. (1979): Modelling and Control of a Gantry Crane, M.Sc. thesis Delft University of Technology (in Dutch).

Hsu, L. and R.R. Costa (1989): 'Variable structure model reference adaptive control using only input and output measurements – part I', *Int. J. of Control*, **49**, pp. 399–416.

Hwang, C.L. and B.S. Chen (1988): 'Adaptive control of optimal model matching in H^∞-norm space', *IEE Proceedings–D*, **135**, pp. 295–301.

Hwang, C.L. and B.S. Chen (1989): 'Model reference adaptive control via the minimization of output error and weighting control input', *IEE Proceedings–D*, **136**, pp. 231–237.

Ioannou, P.A. and K.S. Tsakalis (1986): 'A robust direct adaptive controller', *IEEE Trans. Aut. Control*, **AC-31**, pp. 1033–1043.

Ioannou, P.A. and G. Tao (1989): 'Dominant richness and improvement of performance of robust adaptive control', *Automatica*, **25**, pp. 287–291.

Isidori, A. (1985): *Nonlinear Control Systems: An introduction.* Lecture notes in control and information sciences, Springer-Verlag, Berlin.

Itkis, U. (1976): *Control Systems of Variable Structure*, John Wiley & Sons, New York.

Ionescu, T., and R. Monopoli (1977): 'Discrete model reference adaptive control with an augmented error signal', *Automatica*, **13**, pp. 507–517.

Johnson, C.R. Jr (1980): 'Input matching, error augmentation, self tuning, and output error identification: algorithmic similarities in adaptive model following', *IEEE Trans. Aut. Control*, **AC–25**, pp. 697–703.

Kaufman, H. and C.J. Uliana (1986): 'Model reference and optimal control applied to computerized numerical control systems' *Conf. on Applied Motion Control*, pp. 31–41.

Kiendle, H. (1986): 'Robustheitsanalyse von Regelungssysteme mit Hilfe der Methode der konvexen Zerlegung', *Proc. Aussprachetag Robuste Regelung*, Langen, pp. 19–32.

Kosut, R.L. (1986): 'Methods of averaging for adaptive systems'. In: *Adaptive and Learning Systems: Theory and applications*, pp. 33–45, Plenum, New York.

Kreisselmeier, G. and K.S. Narendra (1982): 'Stable model reference adaptive control in the presence of bounded disturbances', *IEEE Trans. Aut. Control*, **AC-27**, pp. 1169–1175.

Kreisselmeier, G. and B.D.O. Anderson (1986): 'Robust model reference adaptive control', *IEEE Trans. Aut. Control*, **AC-31**, pp. 127–133.

Kwakernaak, H. (1988): 'Robust Control', *Journal A*, **29**, no.4, pp. 17–27.

Landau, Y.D. (1979): *Adaptive Control – the Model Reference Approach.* Marcel Dekker, New York.

Landau, Y.D. and H.M. Silveira (1979): 'A stability theorem with applications to adaptive control', *IEEE Trans. Aut. Control*, **AC-24**, pp. 305–312.

Landau, Y.D. (1980): 'An extension of a stability theorem applicable to adaptive control', *IEEE Trans. Aut. Control*, **AC-25**, pp. 814–817.

Landau, Y.D. and R. Lozano (1981): 'Unification of discrete time explicit model reference adaptive control designs', *Automatica*, **17**, pp. 593–611.

Lee, T.H. and K.S. Narendra (1988): 'Robust adaptive control of discrete-time systems using persistent excitation', *Automatica*, **24**, pp. 781–788.

Lin, Y.H. and K.S. Narendra (1980): 'A new error model for adaptive systems', *IEEE Trans. Aut. Control*, **AC-25**, pp. 585–587.

Lozano, R. and I.D. Landau (1981): 'Redesign of explicit and implicit discrete time model reference adaptive control schemes', *Int. J. of Control*, **33**, pp. 247–268.

Monopoli, R.V. (1974): 'Model reference adaptive control with an augmented error signal', *IEEE Trans. Aut. Control*, **19**, pp. 474–484.

Narendra, K.S. and L.S. Valavani (1978): 'Stable adaptive controller design – Direct control', *IEEE Trans. Aut. Control*, **23**, pp. 570–583.

Narendra, K.S., Y.H. Lin and L.S. Valavani (1980): 'Stable adaptive controller design, Part II: Proof of stability', *IEEE Trans. Aut. Control*, **25**, pp. 440–448.

Narendra, K.S. and Y.H. Lin (1980): 'Stable discrete adaptive control', *IEEE Trans. Aut. Control*, **AC-25**, pp. 456–461

Narendra, K.S. and A.M. Annaswamy (1986a): 'Robust adaptive control using reduced order models', *Proc. 2nd IFAC Workshop on Adaptive Systems in Control and Signal Processing*, Lund, Sweden, pp. 49–53.

Narendra, K.S. and A.M. Annaswamy (1986b): 'Robust adaptive control'. In: *Adaptive and Learning Systems: Theory and Applications*, Plenum, New York, pp. 3–31.

Narendra, K.S. and A.M. Annaswamy (1986c): 'Robust adaptive control in the presence of bounded disturbances', *IEEE Trans. Aut. Control*, **AC-31**, pp. 306–315.

Narendra, K.S. and A.M. Annaswamy (1989): *Stable Adaptive Systems*, Prentice-Hall International, Englewood Cliffs.

Ortega, R. and Y. Tang (1989): 'Robustness of adaptive controllers – a survey', *Automatica*, **25**, pp. 651–677.

Parks, P.C. (1966): 'Lyapunov redesign of model reference adaptive control systems', *IEEE Trans. Aut. Control*, **AC-11**, pp. 362–367.

Parks, P.C. (1985): 'Stability and convergence of adaptive controllers – continuous systems'. In: Self-tuning and adaptive controllers – theory and applications, *IEE Control Engineering Series*, **15**, pp. 93–108.

Passenier, P.O. (1989): An Adaptive Track Predictor for Ships, Ph.D. thesis, Control Laboratory, Delft University of Technology.

Peterson, B.P. and K.S. Narendra (1982): 'Bounded error adaptive control', *IEEE Trans. Aut. Control*, **AC-27**, pp. 1161–1168.

Richalet, J., S. Abu el Ata, L. Delineau, J. Estival and H. Butler (1990): 'Model based predictive control of exotic systems', *Proc. IFAC World Congress*, Tallinn.

Rohrs, C.E., L. Valavani, M. Athans and G. Stein (1985): 'Robustness of continuous-time adaptive control algorithms in the presence of unmodelled dynamics', *IEEE Trans. Aut. Control*, **AC-30**, pp. 881–889.

Sastry, S.S. and A. Isidori (1989): 'Adaptive control of linearizable systems', *IEEE Trans. Aut. Control*, **AC-34**, pp. 1123–1131.

Sastry, S.S. and M. Bodson (1989): *Adaptive Control – Stability, Convergence, and Robustness*, Prentice Hall advanced reference series, Englewood Cliffs.

Slotine, J.-J.E. and Li, W. (1991): *Applied Nonlinear Control*, Prentice Hall International, Englewood Cliffs.

Soeterboek, A.R.M. (1992): *Predictive Control – A Unified Approach*, Prentice Hall International, Englewood Cliffs.

Suzuki, T. and Y. Dohimoto (1978): 'A modified scheme for the Model Reference Adaptive Control with augmented error signal', *Int. J. of Control*, **27**, pp. 199–211.

Suzuki, T. and S. Takashima (1978): 'A hyperstable scheme for discrete model reference adaptive control systems', *Int. J. of Control*, **28**, pp. 245–252.

Tomizuka, M. (1988): 'On a relaxation of SPR condition in parallel MRAS – continuous-time case', *IEEE Trans. Aut. Control*, **AC-33**, pp. 976–978.

Unbehauen, H. (1985): 'Systematic design of discrete model reference adaptive systems'. In: Self-tuning and adaptive controllers – theory and applications, *IEE Control Engineering Series*, **15**, pp. 167–192.

Unbehauen, H. (1987): 'Discrete model reference adaptive systems: design'. In: *Systems & Control Encyclopedia*, ed. M.G. Singh, Pergamon Press, Oxford, pp. 1087–1093.

Verbruggen, H. (1985): *Digital Control Systems*, Delft Universitary Press (in Dutch).

Walrath, C.D. (1984): 'Adaptive bearing friction compensation based on recent knowledge of dynamic friction', *Automatica*, **20**, pp. 717–727.

Index

A

a posteriori error, 113
a priori error, 113
a priori / *a posteriori* correction, 117
  in ITR, 132
  resembling normalization, 115
adaptation gain $\boldsymbol{\Gamma}$, 9, 39
adaptation gains
  choice of, 74
  depending on process gain, 40
  diagonal matrix, 39
adaptation mechanism, 4
adaptive law modification
  dead zone, 91
  if upper bound on parameter is known, 93
  γ-modification, 95
    bounds on γ, 95
    requirements on reference signal, 95
  σ-modification, 94
  pole shift, 94
adaptive laws, 6
  derivation of, 9
  differentiating term, 20
  modification of, 90
  pole shift in, 94, 95
  proportional term, 20, 21
adaptive model adjustment, 190
adaptive offset compensation, 96
augmented error, 58
augmented error method, 50, 52
  auxiliary error, 58
  convergence increase, 71
  derivation of the adaptive laws, 55

derived with hyperstability, 52
derived with Lyapunov's method, 52
design polynomial L, 58
discrete-time, 110
error model, 57
filters L^{-1}, 73
modification of ASGs, 52
parameter vector, 53
primary controller, 52, 53
proportional term in, 52
signal vector, 53
stability proof of, 52
tuning aspects, 73

auxiliary signal generators, 8, 52
freedom in choosing poles of, 73
introducing extra states, 57
linear filters, 54, 62
noise filtering by, 73
replaced by SVFs, 62
transfer function, 64
tuning of, 73

averaging, 22
frequency content in ξ, 25
main results, 25
persistently exciting properties, 25

B

Bernoulli's law, 141
bounded disturbances, 79
bridge
over the gap between theory and practice, 29

C

compensated error, 39, 44
compensator p
making error equation SPR, 39
conditions on process, 7
control signal u
linear in parameters, 8
convergence, 13
convergence rate in adaptive system, 82
convergence requirements, 80
criterion function, 3

D

dead time, 89, 101
problems caused by, 101
solution by orthogonal error signal, 101
dead zone, 91
deadbeat control, 134
deadbeat MRAC, 27
controlling nonminimum-phase processes, 138
definition of error $\epsilon(k)$, 137
practical application, 139
signal and parameter vectors, 136
decomposition model, 159
definition of adaptive control, 2
deterministic input disturbances, 96
digital computers, 109
direct adaptation, 2
direct-drive DC motor, 190
bang–bang control, 195
being a very strong DC motor, 192
choice of reference model, 192
current limitation, 194
current loop, 194
estimation of J_t, 199
experimental results, 202
model of, 194
MRAC compared with PID, 203
nonlinear transfer function, 192
PID optimization, 203
position control, 195
proportional adaptation, 197
reference model, 198
resolver, 196

sensitivity of behaviour to load change, 193
stopping adaptation integrators, 197
time-optimal control, 191, 192, 195
two-sided adaptation, 200
unknown load inertia, 192
discrete MRAC
a priori and *a posteriori* error, 113
augmented error method, 110, 116
ASG choice, 119
ASG reduction, 120
ASGs in, 116
difference from continuous-time, 116
filters L^{-1}, 121
new error model, 118
change from MSTC to RFRC, 134
counterparts of continuous-time designs, 110
hyperstability theory, 112
least-squares adaptation, 123
one-sample delay, 113
stability theory, 110
dominantly rich, 84

E

error augmentation, 51
making error equation SPR, 52
error equation, 12
compensator for, 55
influenced by external disturbances, 79
linear part, 12, 15
made SPR by input modification, 55
nonlinear input term, 38
nonlinear part, 15
error vector e
linear combination of, 44
error-augmenting network, 60
external disturbance, 85
$\nu(t)$, 85
causing parameter drift, 86
depending on reference, 211
using persistent excitation to solve problems caused by, 87

F

fast reference model
requiring wide control boundaries, 191
fictitious reference model, 151

G

gain scheduling, 2
gantry crane scale model, 175
adaptive control of, 182
belt flexibility, 179
control goal, 178
Coulomb friction, 179
mathematical model, 179
model adjustment, 185
model decomposition, 182
practical experiments, 183
virtual mass of trolley, 181
gap
between theory and practice, 6
gradient adaptation, 22

H

hyperstability, 15
design procedure, 19
discrete form of, 112
exchange of requirements, 125
intuitive explanation, 18
modification of, 123
proposal for adaptive laws, 15

I

independent tracking and regulation, 128, 129

a priori / *a posteriori* correction, 132
change from MSTC to RFRC, 134
deadbeat, 135
extension to general MRAC, 133
linear error transfer, 131
minimum-settling time, 135
primary controller, 128
resembling self-tuning, 128
ringing effects, 133
ripple-free response, 135
indirect adaptation, 2
inner loop, 5
input disturbances, 101
input limits, 191
interaction
causing unmodelled dynamics, 217
convergence disturbed by, 215
cross terms caused by, 212
injection in reference model, 215
model adjustment, 212
one-sided, 213
perfect model-matching violation, 217

K

Kalman–Szegö–Popov lemma, 112
Kalman–Yakubovich lemma, 16, 58

L

leakage, 94
least-squares adaptation, 122, 123, 127
forgetting factor, 127
implying time-varying adaptation gain, 127
recursion formula, 127
trace limitation, 128
trace scaling, 128
violating passivity requirement, 123
violating SPR property, 110
limit cycles, 20, 89
Lyapunov function, 12, 39
choice of, 12
integral properties, 80
quadratic function of e and ϕ, 39
taking derivative of, 39
time derivative of, 13
Lyapunov stability, 236
Lyapunov's equation, 13, 16, 39, 40
Lyapunov's method, 12, 38
discrete form of, 111
reconsidered, 80
Lyapunov's theorem, 13

M

measurement noise, 77, 79, 85
memory in adaptation, 20
minimum-settling-time control, 134
model adjustment
adaptive, 191
application, 217
direct, 188, 212
model reduction, 28
model state
reconstruction created by SVFs, 51
use as feedforward, 51
monic, 64
MRAC
complete state information, 35
continuous-time, 34
definition of, 6
development of, 6
discrete-time approach, 109
heart of, 6
multi-input, multi-output, 44
parallel, 5
practical applications, 6
recent issues, 28
series, 5
series–parallel, 5
using output feedback, 50

multi-input, multi-output processes, 210
multi-input systems, 44
MUSIC, 139, 183, 202, 224

N

nominal process, 78
nonlinear extra dynamics, 3
nonlinear systems, 28
normalization, 20

O

obscure feedback loop, 2
one-sample delay, 113
  phase shift caused by, 115
optimal control, 27
orthogonal error, 103, 104
  calculation procedure, 103
  determination of, 103
  incompatibility between values, 104
  weighting factor α, 104
outer loop, 5

P

parallel state reconstruction, 50
parameter drift, 86, 87, 91
parameter error, 12
passivity, 239, 240
passivity and PR
  similarity between, 17, 240
perfect model matching, 7, 35, 36, 50
  ASG1 and ASG2 modification, 66
  ASG2 modification, 66
  in discrete augmented error method, 116
  with modified ASGs, 63
  without ASG modification, 63
persistent excitation
  definitions of, 81
  degree μ_0 of, 82
  degree m of, 82
  for continuous-time signals, 82
  for discrete-time signals, 81
  guaranteeing UAS of parameter error, 81
  link between definitions, 82
  link with Lyapunov, 80
  requirements in the presence of disturbances, 87
  spectral lines, 82
  threshold on μ_0, 87
persistently exciting, 13, 80, 81
perturbed systems
  stability properties of, 84
phase-variable form, 35, 36, 44, 45, 63
Popov's integral inequality, 17
positive real, 239
positivity, 239
practical application
  direct-drive DC motor, 190
  gantry crane scale model, 175
  propeller setup, 217
  water-level system, 139
predictive adaptive control, 5
predictive control, 3
primary controller, 4, 6, 7
primary feedback, 2
process dead zone, 89
process input limitations, 89
process nonlinearities, 89
propeller setup, 217
  disturbance compensation in process, 228
  experiments, 224
  interaction in, 221
  mathematical model, 219
  mutual interaction, 225
  nonlinearity compensation, 222
  one-sided interaction, 225
  primary controller, 222
  reference model, 223
proportional adaptation, 21

R

reference generator, 5
reference model, 4, 6, 8
  cautious, 8
  choice of, 8
  controllability, 8
  minimum-phase, 8
  perfect model matching, 8
  relative degree, 8
  stability, 8
reference model adjustment, 215
reference model decomposition, 155
  adjustment to process capabilities, 188
  allowing simple primary controller, 156
  basic idea, 157
  changing control goal, 164
  choice of design polynomials, 163
  design polynomials T_1 T_2 and N, 159
  design steps, 174
  differences from predictive control, 156
  effects on error equation, 159, 160, 162
  effects on output disturbance, 161
  example, 164, 169
  freedom in choice of N, 175
  nominal part, 155
  originating from predictive control, 156
  performance price to be paid, 164
  practical application, 175
  reduced control effort, 188
  structured unmodelled part, 155
  to use low-order controller, 155
  unmodelled part, 155
relatively prime, 64
resolver, 196
RFRC MRAC
  controlling nonminimum-phase processes, 138
  definition of error $\epsilon(k)$, 137
  practical application, 139
  signal and parameter vectors, 136
ripple-free control, 134
robust control, 26
  and adaptive control, 78
  controller design, 78
  linked with sensitivity analysis, 78
  with the aid of Lyapunov's method, 78
robustness, 78
  from a nonadaptive point of view, 78
  in adaptive context, 78, 79
  performance, 78
  stability, 78
  state-dependent disturbances, 151
  unmodelled dynamics, 151
robustness in MRAC
  analysis tools, 79
  experiment to test, 90
  improvement by adaptive law modification, 90

S

saturation, 5
secondary feedback, 2
self-tuning control, 5, 28
  unification with MRAC, 28
sensitivity approach, 9
sensitivity model, 10
  implementation, 10
signal error, 12
slow drift instability, 150
small gain theorem, 19
state feedback, 35
state-dependent disturbances, 148

state-variable filters, 50
strict passivity, 240
strictly positive real, 16, 239
sufficiently rich, 83
superpassivity, 19
superstrictly positive real, 19
system class $L(\boldsymbol{\Lambda})$, 123
system class $N(\boldsymbol{\Pi})$, 125
system noise, 77, 79, 85

T

total stability, 84
    link with persistent excitation, 84
tracking and regulation, 5
trajectory generator, 5

U

unmodelled dynamics, 148
    adaptive law modification, 154
    causing instability, 149
    definition of $\boldsymbol{\theta}^*$, 151
    description of, 151
    disturbance caused by, 152
    disturbing error equation, 152
    fictitious reference model, 151
    instability mechanisms, 149
    output disturbance caused by, 154
    parameter drift causing instability, 149
    PE properties, 154
    SPR property violated by, 154
    using modifications to cope with, 154
    violating perfect model matching, 150

V

variable structure systems, 20, 27
    resemblance with MRAC, 27

W

water-level system, 139
    block diagram, 143
    description of, 139
    modelling of, 142
    practical results with RFRC MRAC, 144
Whitacker, 4
wind up, 20